W9-AXJ-539

ANIMAL PAIN
Perception and Alleviation

S. R. Geiger, *Publications Manager and Executive Editor*
B. B. Rauner, *Production Manager*
O. Lowy, L. S. Chambers, S. Mann, *Editorial Staff*
C. J. Gillespie, *Indexer*

ANIMAL PAIN

Perception and Alleviation

EDITORS

Ralph L. Kitchell
Department of Anatomy, School of Veterinary Medicine, University of California, Davis, California

Howard H. Erickson
Department of Anatomy and Physiology, College of Veterinary Medicine, Kansas State University, Manhattan, Kansas

ASSOCIATE EDITORS

E. Carstens
Department of Animal Physiology, University of California, Davis, California

Lloyd E. Davis
Clinical Pharmacology Studies Unit, College of Veterinary Medicine, University of Illinois, Urbana, Illinois

AMERICAN PHYSIOLOGICAL SOCIETY
Bethesda, Maryland

Library of Congress Catalog Card Number 83-15867

International Standard Book Number 0-683-04625-X

Printed in the United States of America by
Waverly Press, Inc., Baltimore, Maryland 21202

Distributed for the American Physiological Society by
The Williams & Wilkins Company, Baltimore, Maryland 21202

Preface

Pain is a complex physiological phenomenon; it is hard to define satisfactorily in human beings, and it is extremely difficult to recognize and interpret in animals. Scientific knowledge concerning pain perception in animals must be obtained by drawing analogies based on comparative anatomy, physiology, and pathology and by inference based on subjective responses to pain experienced by humans. Debate continues about whether animals of different species perceive pain similarly and whether any species perceives pain the same way humans do. The use of animals in research, in education, and in testing products to minimize adverse effects requires more knowledge about pain perception in animals. Increasing public concern about animal welfare has added urgency to this need.

Our knowledge of the scientific basis of the mechanisms of pain has advanced substantially in the last two decades. Nociceptors, or pain receptors, are widespread in the skin and tissues of animals; chemical mediation of nociceptor excitation may provide a key to understanding the peripheral phenomena related to pain. The expression of pain in animals involves multiple ascending and descending branches as well as specialized pain-signaling mechanisms in the spinal cord. The importance of these different pathways varies with species and circumstances. Endogenous neural systems in the brain stem and forebrain, including both opioid and nonopioid mechanisms, may modulate the central transmission of nociceptive signals in animals. Noxious stimuli mediate a variety of different functions; each animal has a consistent response to noxious stimuli or a consistent pattern of escape from pain.

A symposium on pain in animals was held at the 66th Annual Meeting of the Federation of the American Societies for Experimental Biology (FASEB) in New Orleans on April 20–21, 1982. Other workshops and meetings on the subject of pain have focused primarily on pain in humans; this is the first symposium known to us that has specifically addressed pain in animals. The American Veterinary Medical Association (AVMA) Council on Research, with financial support from the AVMA Foundation, planned and organized the symposium.

Cosponsors were the American Physiological Society and the American Society for Pharmacology and Experimental Therapeutics, constituent members of FASEB, who contributed suggestions on planning and organization.

The goal of the symposium was to establish the factual background on acute pain in animals from which more detailed and specific information can be developed. Chronic pain in animals was not specifically addressed, but the symposium did project research needs to better measure and alleviate pain, discomfort, and anxiety in experimental animals. As the mechanisms of pain are better understood, the humane treatment and alleviation of pain in experimental animals can be placed on a much firmer scientific basis.

The book is divided into two sections. Pain perception in animals focuses on peripheral and supraspinal mechanisms involved in pain, segmental neurophysiological mechanisms, spinal cord pathways and control systems, stimulation analgesia and endorphins, behavioral mechanisms for assessment of pain, assessment of pain during surgical procedures, and phylogenetic evolution of pain expression in animals.

Alleviation of pain in animals covers drug-disposition factors and species variation, evaluation of analgesic drugs in horses, and control of pain in dogs and cats. There is a significant variation among species in the absorption and biotransformation of drugs used to alleviate pain in animals.

This book is intended as a source of basic information about the perception and alleviation of pain in animals for scientific investigators working in this area; for veterinarians interested in the health and welfare of animals, the assessment of pain during surgery, and the alleviation of pain; and for individuals involved in federal, state, and local regulation of the use of animals in research, education, and quality control of human and animal health products.

As chairman of the planning committee for the symposium, I thank the session chairmen, speakers, members of the planning committee (L. E. McDonald and P. K. Hildebrandt), other members of the AVMA Council on Research, R. L. Kitchell (consultant to the Council), and E. R. Ames (AVMA staff consultant).

We are grateful to the AVMA Foundation for financial support of the symposium. We thank the American Physiological Society, O. E. Reynolds, S. R. Geiger, editors, and staff for assistance in meeting arrangements and the publication of this book.

Howard H. Erickson

Introduction: What is Pain?

Pain in animals is a topic of considerable interest and debate involving strong human emotions affected by differing personal beliefs of individuals with widely divergent backgrounds. Any discussion of pain involves semantics. Pain has been defined by different individuals based on personal experiences. Most authorities agree that pain is a perception, not a physical entity, and that perception of pain depends on a functioning cerebral cortex. Unlike most other sensations, no single area of the cerebral cortex seems specifically necessary for the perception of pain. It is semantically incorrect to refer to "pain" stimuli, impulses, pathways, or reflexes because these occur in humans, and presumably in animals, without perception of pain. The term *noxious* describes stimuli that, if perceived, give rise to the perception of pain. Sherrington (3) originally defined noxious as a stimulus actually or potentially damaging to the skin. The receptors specifically responsive to noxious stimuli are termed *nociceptors*. A stimulus must be a certain strength before a nociceptor will generate nerve impulses in the peripheral nerve fiber of which it is a part. This stimulation strength is called the *nociceptive threshold*. In certain circumstances this amount of neural activity may be too little to result in perception of pain. The strength at which noxious stimulation is perceived by a human being as pain is referred to as the *pain detection threshold*. The strongest intensity of noxious stimulation that a human being will permit an experimenter to deliver is called the *pain tolerance threshold*. The strength of noxious stimulation necessary to reach the nociceptor threshold is rather constant and varies little among humans and animals. The strength needed to cross the pain detection threshold is slightly more variable, especially among humans experiencing clinical pain (4). The pain tolerance threshold is the most variable of the three thresholds. Most clinical veterinary neurologists are amazed by the high pain tolerance thresholds of some dogs.

Introspective verbal reports are the most frequently used method of assessment of pain perception in human beings. Some authorities have developed objective measurements (e.g., manipulation of a lever or drawing lines of varying length) to better quantitate pain detection and pain tolerance thresholds. When considering pain in animals, analogies must be drawn between human and animal anatomy, physiology, and behavior. Knowledge about pain in animals remains inferential, however, and neglect of the probabilistic nature of pain perception in animals leads to anthropomorphism (1). On the other hand, overemphasis on the uncertainty of our knowledge about pain perception in animals, which leads to a denial that pain perception exists in animals, is logically as well as empirically unfounded (1). The tacit assumption is that stimuli are noxious and strong enough to give rise to the perception of pain in animals if the stimuli are detected as pain by human beings, if they at least approach or exceed tissue-damaging proportions, and if they produce escape behavior in animals (2).

This book describes the neuromechanisms involved in pain perception and alleviation in animals. It is hoped that this volume will inspire additional studies and symposia dedicated to developing better methods for the assessment of pain and its prevention and alleviation.

REFERENCES

1. Kitchell, R. L., Y. Naitoh, J. E. Breazile, and J. M. Lagerwerff. Methodological considerations for assessment of pain perception in animals. In: *The Assessment of Pain in Man and Animals*, edited by C. A. Keele and R. Smith. London: Universities Federation for Animal Welfare, 1962, p. 244–261.
2. Lineberry, C. G. Laboratory animals in pain research. In: *Methods of Animal Experimentation*, edited by W. I. Gay. New York: Academic, 1981, vol. 6, p. 237–311.
3. Sherrington, C. *The Integrative Action of the Nervous System*. New Haven, CT: Yale Univ. Press, 1947, p. 228.
4. Sternbach, R. A. *Pain: A Psychophysiological Analysis*. New York: Academic, 1968, p. 18–26.

Ralph L. Kitchell
Howard H. Erickson

Contents

Alleviation of Pain

ONE

Peripheral Mechanisms Involved in Pain

Lawrence Kruger
Barbara E. Rodin

Departments of Anatomy and Anesthesiology, and the Ahmanson Laboratory of Neurobiology, Brain Research Institute, University of California Center for Health Sciences, Los Angeles, California

Specificity Versus Pattern Theory • Anatomy and Physiology of Nociceptors: Small myelinated afferents, Unmyelinated afferents, Chemical mediation • **Pathology Associated With Nerve Injury:** Peripheral sprouting of nociceptive afferents in noninjured nerves, Neuroma formation

The structural and physiological basis of pain sensations has been the most elusive area of sensory research, and the information concerning specific receptors and pathways is mostly recent. The nineteenth century concept of sensory modality, introduced by Helmholtz to designate qualitatively distinctive sensory continua, was widely adopted to fit a schema in which touch, warmth, cold, and pain were accepted as primary qualities or modalities (10, 68, 107) in keeping with the Müller concept of specific energy. Müller's concept is more easily applied to the organs of special sense than to cutaneous sensation because it is often difficult to excite the skin and evoke a single introspective sensory quality; for example, it is difficult to evoke pain without touch or warmth components. The discovery by Blix (8) of a mosaic of sensory-specific spots ultimately led to acceptance of the idea that there was a distinctive anatomical substrate subserving each specific modality. Von Frey allocated the end organs known in the nineteenth century to each of the established modalities (68, 98, 107).

Specificity Versus Pattern Theory

The von Frey theory has been continually challenged on numerous grounds. The separability of cold and warm spots seemed clear enough, but the density of touch and pain even led Blix to doubt their spatial distinctiveness. The von Frey assignment of Krause end bulbs to cold sensations and Ruffini endings to warm sensations has been challenged by modern knowledge, but his assignment of free nerve endings to pain is still widely accepted, and his association of hair follicle basket formations and Meissner corpuscles to touch in hairy and glabrous skin, respectively, is almost certainly a correct, if incomplete, correlation. Goldscheider (39) challenged the problem of specificity with particular reference to pain, because pain could be evoked by a variety of stimuli if they were sufficiently intense. The concept of pain as an affective component distinct from a sensory modality has survived into the modern era, although it has long been recognized that there are conditions (e.g., syringomyelia, cocaine and nitrous oxide analgesia) in which an absence of pain is not accompanied by loss of other sensory qualities.

The introduction of electrophysiological techniques helped resolve some of the conflicts, and Adrian's pronouncement of the "all-or-none law" in peripheral nerve axons, followed by the pressure and anesthetic nerve-block experiments of Gasser and Erlanger, provided new support for the Sherringtonian doctrine that a specific adequate stimulus could be determined for each peripheral axon. In the 1950s several workers defined the adequate stimulus for axons with pain endings in terms of their clearly elevated thresholds to all stimuli (54, 55, 59, 81). However, many investigators uphold the view that each receptor responds to a range of stimuli, which it converts into patterns of axonal impulses rather than into specific-line modality information, and that the discharge pattern from several receptors is essential for sensory discrimination (82, 83).

In the following decade the laboratories of Iggo and Perl established the existence of coherent populations of visceral and cutaneous receptors with unmyelinated or thinly myelinated axons that possess elevated thresholds for innocuous stimuli and that discharge vigorously only in response to noxious stimulus intensities (16). The existence of nociceptive cutaneous sense organs has been confirmed in the past decade for many species, including humans, and the features of myelinated and unmyelinated classes have been characterized in several laboratories (107). These data are consistent with human experiments demonstrating that electrical stimulation of slowly conducting myelinated fibers elicits fast, pricking pain; slow, burning pain is elicited in unmyelinated fibers when repetitively stimulated (7, 28). Although there is no report of unpleasantness or pain until the thinly myelinated fibers are excited (28, 42), stimulation of only the faster A fibers correlated with tactile sensation can produce both pain and nociceptive reflexes (127) when excited repetitively; a few fast-conducting myelinated fibers appear to display high thresholds (51). Recent evidence relating individual fibers to specific sensations is derived from the observations of Torebjörk and Ochoa (116) in which microelectrode recording from single fibers in humans revealed that the natural adequate stimulus for elicting discharge in each fiber correlated accurately with the sensation aroused by electrically stimulating the same fiber. A crucial observation is that excitement of only one myelinated or unmyelinated nociceptor axon is necessary to produce pain sensation without invoking a complex pattern of multiple inputs.

Anatomy and Physiology of Nociceptors

The prevailing nineteenth century view that the morphological substrate of pain can be assigned to free nerve endings still survives largely because of their ubiquitous distribution and especially because pain can be elicited from zones of hairy skin lacking the corpuscular receptors commonly found in glabrous skin (108). The terms *free*, *bare*, and *naked* were originally derived principally from studies with silver impregnation and, to a limited extent, vital staining with methylene blue. The deficiencies of both methods have been reviewed by numerous authors (41, 89, 107). It is generally recognized that metallic impregnation is often erratic and incomplete in delineating unmyelinated fibers and additionally is subject to artifacts of overimpregnation, including staining of collagen and reticular fibers. Methylene blue provides more specific axonal visualization but fails to reveal many of the endings that are apparent with acetylcholinesterase histochemistry (29). Our current investigations, involving the axonal transport of a horseradish peroxidase–lectin conjugate, suggest that both metallic impregnation and vital stains provide an incomplete picture. In any case, these light-microscopic methods do not even reveal the envelopment of axons by Schwann cells. Several authors have suggested that the term *free nerve ending* is misleading and should be discarded because it implies that the axon is bare, a condition that does not exist when these endings are examined with the electron microscope (20, 69, 94). It is now recognized that a complete description of a cutaneous nerve ending and its surround can only be achieved through electron-microscopic examination.

The limitation of morphological studies makes correlating structure and function difficult, but identification of physiologically characterized receptive-field spots has been achieved with electron-microscopic methods for the description of the Merkel cell complex of Pinkus-Iggo domes (57) and cold spots (47). A similar approach seemed applicable to one class of fibers implicated in pain (69).

Small myelinated afferents

Cutaneous mechanical nociceptors (also known as high-threshold mechanoreceptors), with afferent fibers conducting between 5 and 50 m/s, constitute a functionally congruent class in the hindlimb nerves that innervate the hairy skin of cats and monkeys (15, 97). Their average conduction velocity of under 30 m/s (Fig. 1) corresponds to the fastest component of the δ-wave of the compound action potential. This feature has led to calling these afferent units δ-nociceptors, although those with lowest threshold and highest conduction velocity may extend slightly beyond the δ-nociceptor range.

The receptive field of each myelinated nociceptor afferent unit consists of an array of spots, each causing a response when indented by a strong mechanical stimulus (Fig. 2). Each excitable point (usually $<250\ \mu m$) is surrounded by an inexcitable region from which responses cannot be elicited with stimulus intensities several times greater than those sufficient to activate the responsive spots (Fig. 3). Mapping the discontinuous nature of the receptive field requires the use of small, sharp probes and avoidance of skin penetration and damage to prevent receptor inactivation. Repeated stimulation may also enhance their responsiveness (4, 18, 35), a process commonly referred to as sensitization.

A series of 17 such spots, mapped and bracketed by the insertion of stainless steel insect pins, was studied in the distribution of the posterior femoral cutaneous nerve of the cat hindlimb. A search was made for an isolated thinly myelinated axon in semithin plastic sec-

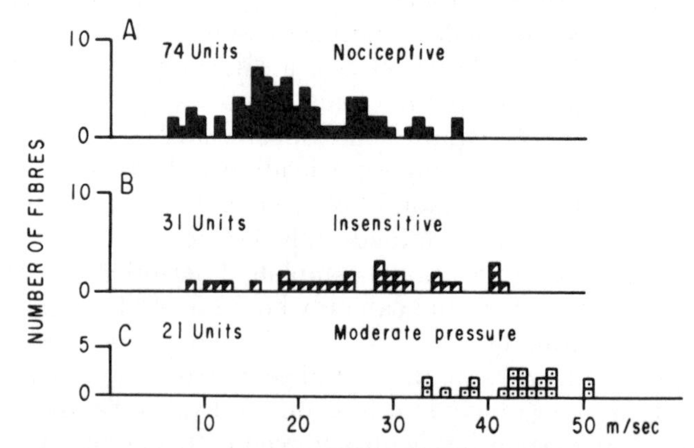

Fig. 1. Conduction velocities of high-threshold mechanoreceptor fibers in cat hindlimb. Those fibers with highest threshold in nociceptive range tend to cluster at low range of conduction velocity; those requiring moderate pressure tend to cluster at high range of δ-fibers and above. [From Burgess and Perl (15).]

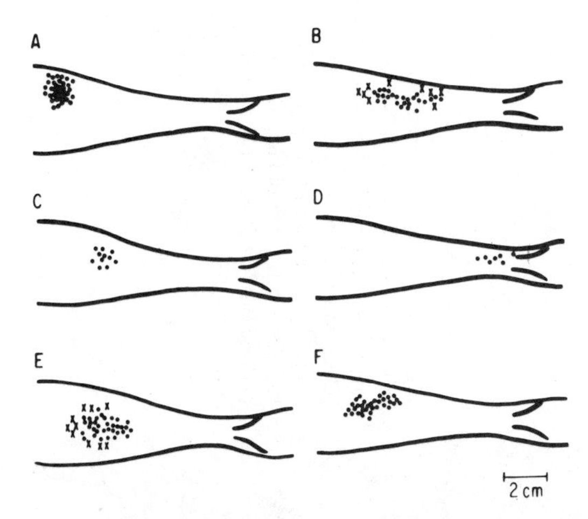

Fig. 2. Receptive fields of 6 thinly myelinated nociceptor fibers innervating cat hindlimb. Each receptive field consists of a variable cluster of spotlike zones surrounded by insensitive domains. Conduction velocities 9.5–29 m/s. [From Burgess and Perl (15).]

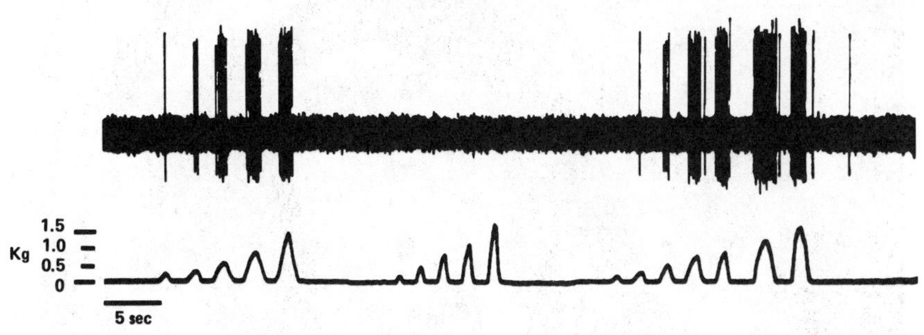

Fig. 3. Discharge from 2 adjacent spots (5 mm apart) of a 17-spot receptive field recorded from myelinated nociceptor fiber in cat (*upper trace*), demonstrating discontinuous nature of receptive field. Force applied with 1-mm probe (*lower trace*) to 1 spot (*upper left*) reveals unresponsive intermediate zone (*middle*) and similar response at nearby spot (*right*). Force values are reduced with use of sharp probe but with risk of receptor damage. [From Kruger, Perl, and Sedivec (69).]

tions of each block, with two pinholes used for orientation. Subsequent semithin and thin sections were traced by following an axon leaving a mixed bundle in its upward course into the dermal papillary layer within the marked zone (Fig. 4). The distal unmyelinated continuation of a thinly myelinated axon into one of the papillary ridges within the demarcated zone was consistently found in each marked spot. Although the dermal papillae within this zone contained other specialized structures, the most consistent feature of each block examined was the presence of several unmyelinated-axon profiles surrounded

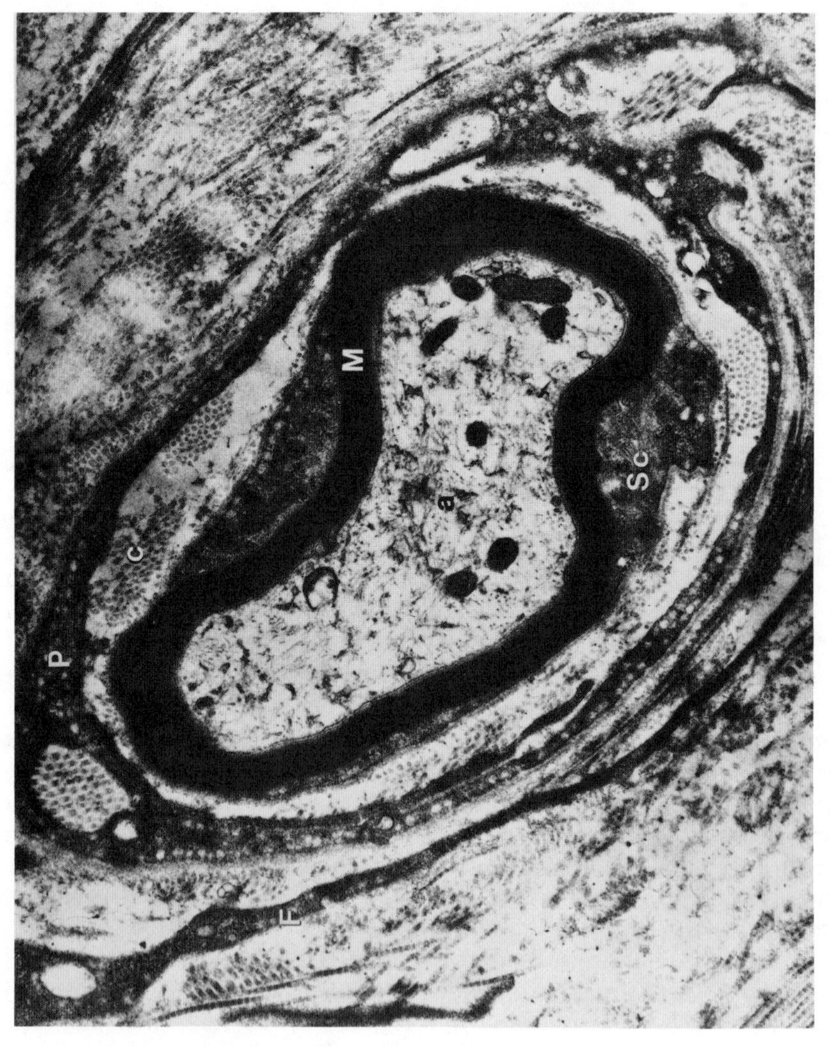

Fig. 4. Electron micrograph of isolated small myelinated (*M*) axon (*a*) and its Schwann cell (*Sc*) in subpapillary dermis below marked nociceptor spot, surrounded by perineurial sheath (*P*), fibroblasts (*F*), and endoneurial collagen (*c*). ×12,700. [From Kruger, Perl, and Sedivec (69).]

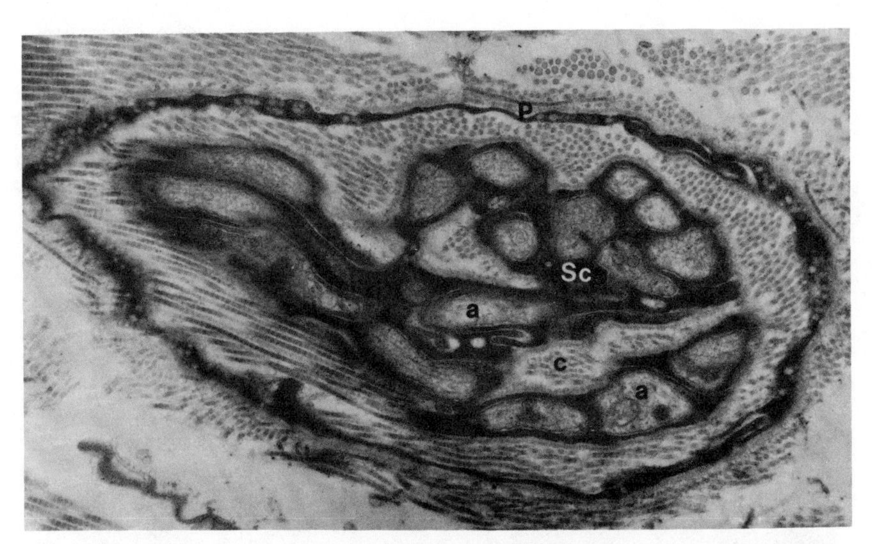

Fig. 5. Electron micrograph of typical dermal unmyelinated nerve bundle with numerous axons (*a*) surrounded by single Schwann cell (Sc), endoneurial collagen (c), and thin layer of perineurium (*P*). ×11,800. [From Kruger, Perl, and Sedivec (69).]

by a thin layer of Schwann cell cytoplasm and its basal lamina (Fig. 5). The several profiles may belong to a single undulating fiber, its branches, or a bundle of separate axons, but their failure to disperse widely suggests that this arrangement is a feature characteristic of myelinated mechanical nociceptors.

The axon–Schwann cell complex was traced to the epidermal border where the thin Schwann cell basal lamina merges with the thicker, denser basal lamina underlying the basal keratinocytes (Fig. 6). Near the juncture of the dense matrix of Schwann cell processes interdigitating with the ribosome-rich keratinocytes, the Schwann cytoplasm often displayed clusters of micropinocytotic vesicles. As the axon penetrated the epidermis it was accompanied by thin Schwann cell processes until it was completely enveloped by a basal keratinocyte (Fig. 7), a feature that is apparently absent in other receptor arrangements and was not regularly observed outside of the marked zone in the same specimens. Within the axons, at the site of epidermal penetration, there were sparse clusters of clear round or pleomorphic vesicles and occasional large dense-core vesicles, but these were rare after the axon–Schwann cell complex was enveloped within the epidermis (Fig. 7) or when the axon lost its Schwann sheath and was completely surrounded by keratinocytes (Fig. 8).

The intraepidermal axon is clearly never a free or bare nerve ending, nor is the nature of its receptive site ever certain. Whether any axons can be traced to more superficial layers is dubious (87), since no

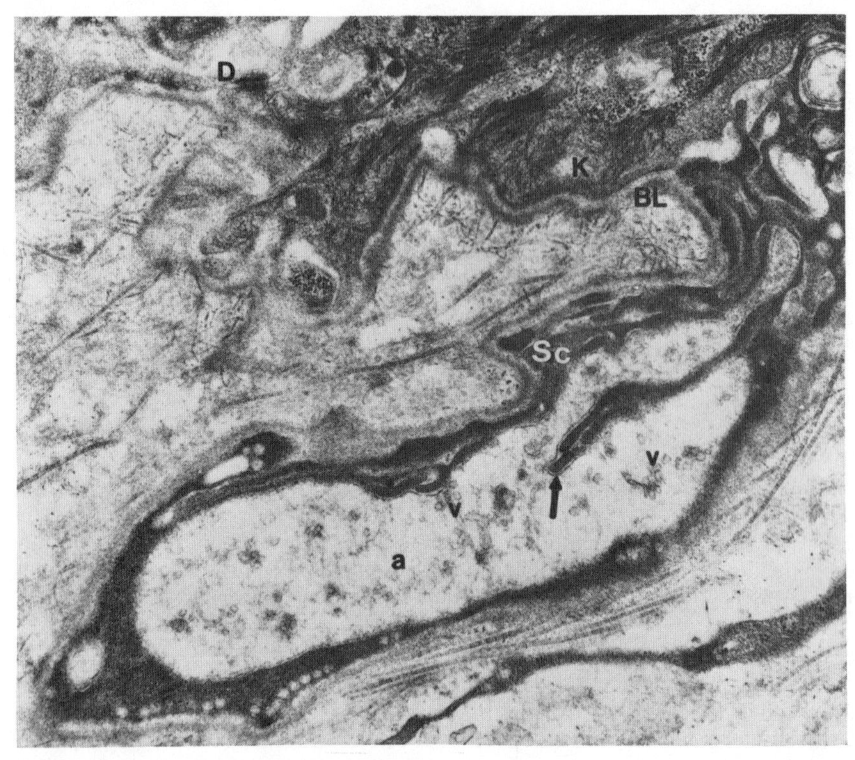

Fig. 6. Electron micrograph of epidermal penetration site of single unmyelinated axon (*a*) containing patches of clear vesicles (*v*) and a few dense-core vesicles, surrounded by Schwann cell processes (*Sc*) that appear to bifurcate (*arrow*) near the axon's contact with basal lamina (*BL*) of epidermal keratocyte (*K*), displaying typical desmosomal junction (*D*). Note fusion of basal laminae and Schwann cell pinocytotic vesicles in marked nociceptor spot. ×12,800. [From Kruger, Perl, and Sedivec (69).]

convincing examples of intraepidermal axons unassociated with Merkel cells have been observed in several decades of electron-microscopic study (11, 12). This is somewhat difficult to reconcile with silver-impregnated fibers, some having been traced into the stratum corneum with optical microscopy (Fig. 9; 41, 106). However, there is some recent evidence of deep intraepithelial axons in electron micrographs of the oral mucosa (Y. Yeh and M. R. Byers, personal communication), and it is conceivable that the widely spaced intraepidermal axons seen with metallic impregnation might correspond to nociceptor spots.

To date, in each example of morphological correlation, examination of a spotlike receptive field has revealed an axon penetrating the epidermal basal lamina; this now appears to be the rule for Merkel disks (57), cold spots (47), and myelinated nociceptors (69). Because most other afferent fibers display broadly distributed receptive fields,

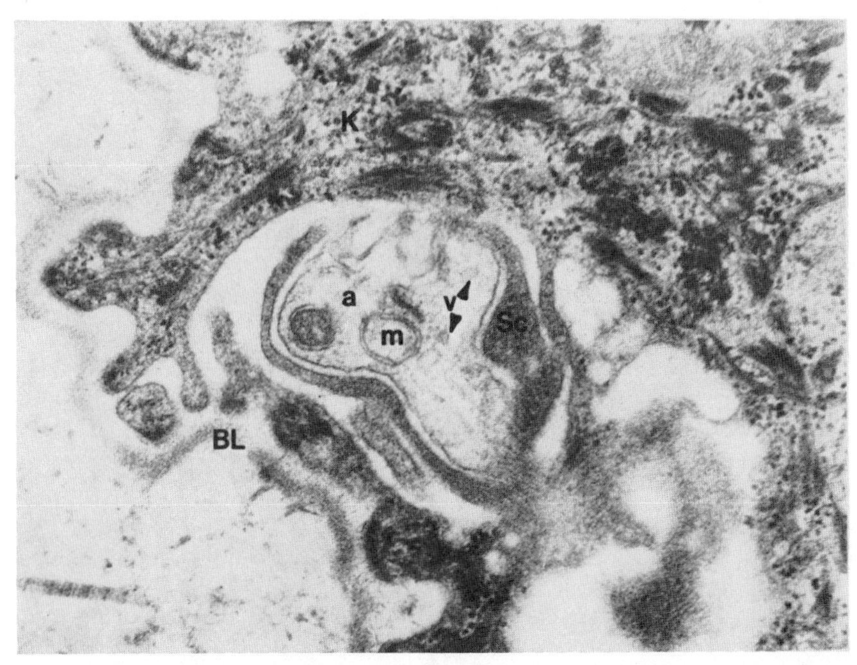

Fig. 7. Electron micrograph of intraepidermal axon (*a*) containing mitochondria (*m*) and a few clear vesicles (*arrowheads*) enveloped by Schwann cell cytoplasm (*Sc*) lacking basal lamina (*BL*) of surrounding keratocyte (*K*). ×38,600. [From Kruger, Perl, and Sedivec (69).]

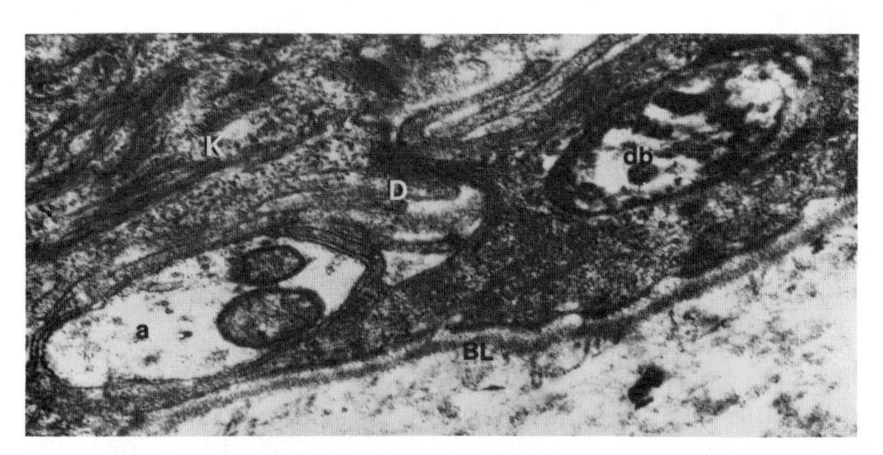

Fig. 8. Electron micrograph of a bare intraepidermal axon (*a*) in marked nociceptor spot completely surrounded by keratocytes (*K*) with typical desmosomal junction (*D*) and dense body (*db*) at border of epidermal basal lamina (*BL*). ×36,600. [From Kruger, Perl, and Sedivec (69).]

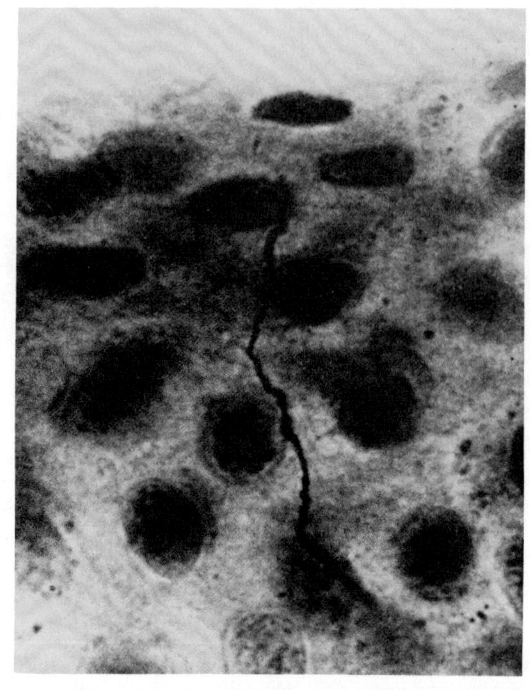

Fig. 9. Silver-impregnated human intraepidermal fiber approaching epidermal stratum corneum. [From Ridley (102a), by permission of Cambridge Univ. Press.]

it is tempting to speculate that their terminals remain within the dermis, since the weight of evidence already supports this conjecture.

Unmyelinated afferents

The broad distribution of bundles of unmyelinated (C) axons constituting the largest proportion of limb nerves represents some of the most perplexing unresolved problems in understanding cutaneous sensation. A large proportion of the C-fiber population innervating proximal hairy skin consists of sensitive mechanoreceptors (3), but these appear to be rare in distal nerves innervating the glabrous skin of primates, including humans (70, 114, 118), where polymodal nociceptors appear to be the dominant population.

A uniform classification of cutaneous unmyelinated nociceptors is difficult to devise because *1*) recovery after a noxious mechanical, thermal, or chemical stimulus is slow; *2*) there is always the danger of receptor inactivation resulting from damage; and *3*) it is difficult to record from single fibers for prolonged periods. Consequently the description of each unit is often incomplete. Nevertheless there is clearly some variety in this population.

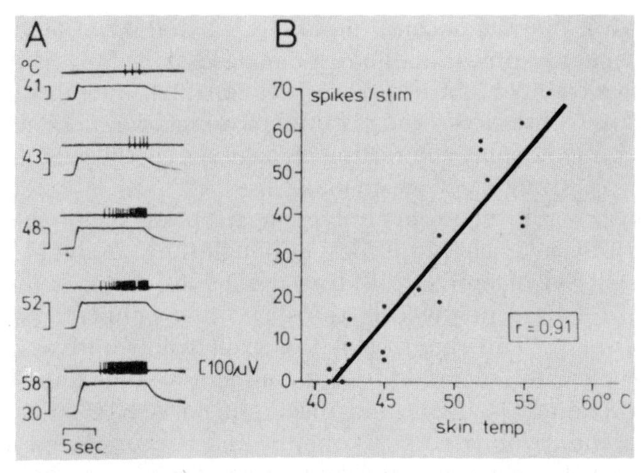

Fig. 10. Responses of polymodal nociceptor C fiber to graded noxious thermal stimuli showing linear relationship between discharge rate and cutaneous temperature. [From Beck, Handwerker, and Zimmermann (2).]

Mechanical C nociceptors that respond to strong mechanical displacement with a low-frequency slowly adapting regular discharge appear to require distortion of the deep subcutaneous fatty layer (4, 56); provided the skin remains unbroken, this type does not respond to a wide range of thermal stimuli or to irritant chemicals (16). Many mechanical C nociceptors respond even more vigorously to noxious heat (Fig. 10), a property also displayed by many myelinated mechanical nociceptors (1, 16, 18, 35, 37, 58, 70, 71, 114, 118). The mechanical-thermal C nociceptors also appear to be deep, sometimes respond sluggishly to cold, and possess small oval receptive fields frequently found near a subcutaneous vein (16). Most C nociceptors are generally classified as polymodal, possess small receptive fields, are excited by more superficial mechanical stimuli, and have elevated but not extreme thresholds. They differ from the above categories by their prompt rapidly or slowly adapting response to noxious heat (84) and their usual sensitivity to chemical stimuli applied to the unbroken skin surface (4, 16, 70). Further subclassification based on differential sensitivity to chemical agents, such as bradykinin and serotonin, may also be warranted (1, 129). Polymodal nociceptors frequently display sensitization with repeated stimuli, although in both C fibers and δ-fibers this effect may differ in hairy and glabrous skin (19). Thresholds to heat stimuli can be lowered by intra-arterial injection of chemical substances derived from chemical damage (99, 100, 129); Perl (100) suggested that the liberation of polypeptides, histamine, serotonin, and prostaglandins may sensitize the terminals so that they can signal weak stimuli, thereby accounting for the elicitation of pain by relatively innocuous stimuli applied to damaged skin. Surprisingly, vascular occlusion does not alter sensitization (80), and the nature of putative pain substances remains undetermined. Polymodal nociceptive units rarely display the resting discharge hypothesized to be held in check by a gating mechanism (83).

The sensory role of unmyelinated nociceptors might be inferred from their physiological properties; although they conduct too slowly to account for nociceptive withdrawal reflexes, there is ample evidence that they are involved in the slow, intense burning quality of pain experience. The security of this conclusion derives from the circumstance that separation of myelinated- and unmyelinated-fiber conduction by ischemic and anesthetic nerve block is easier than untangling the myelinated components, and because there are some locations (e.g., the cornea) that are innervated solely by unmyelinated axons (107). However, although there is the long-accepted correlation between C fibers and a variety of painful experiences and the noxious-stimulus properties required for their excitation, there are clearly many unmyelinated fibers that are sensitive to delicate mechanical or small thermal changes. This circumstance and the widespread distribution of unmyelinated autonomic fibers poses a serious problem for the morphologist attempting to determine the nature of unmyelinated pain endings.

There are numerous descriptions of deep and subpapillary dermal nerve nets of unmyelinated fibers, sometimes termed *plexus* because of their extensive interdigitation. Unmyelinated fibers are numerous in bundles containing myelinated fibers, but there are also purely unmyelinated bundles (see Fig. 5). These bundles ultimately discard their perineurial sheath and splay out in a fashion consistent with the descriptions of free nerve endings derived from optical microscopy (Fig. 11); although some profiles contain a few vesicles and mitochondria, it is uncertain which structures should be designated *end organ*. Unmyelinated fibers constitute the majority of terminal structures and can be traced throughout the dermis to the dermal-epidermal boundary, hair follicles, Pacinian corpuscles, and blood vessels. The most extensive electron-microscopic studies of these fibers by Cauna (21–23) provide a description of many of these patterns.

In his study of human glabrous skin, Cauna (23) described three types of free nerve endings: open, beaded, and plain. Plain endings can be traced to the epidermal border, each indenting the epidermis in a punctate fashion, in contrast to the broad horizontal extent of the highly branched free penicillate endings found in human hairy skin (Fig. 12; 21). The problem in interpreting Cauna's findings, as well as those of all other authors, is the lack of suitable criteria for determining

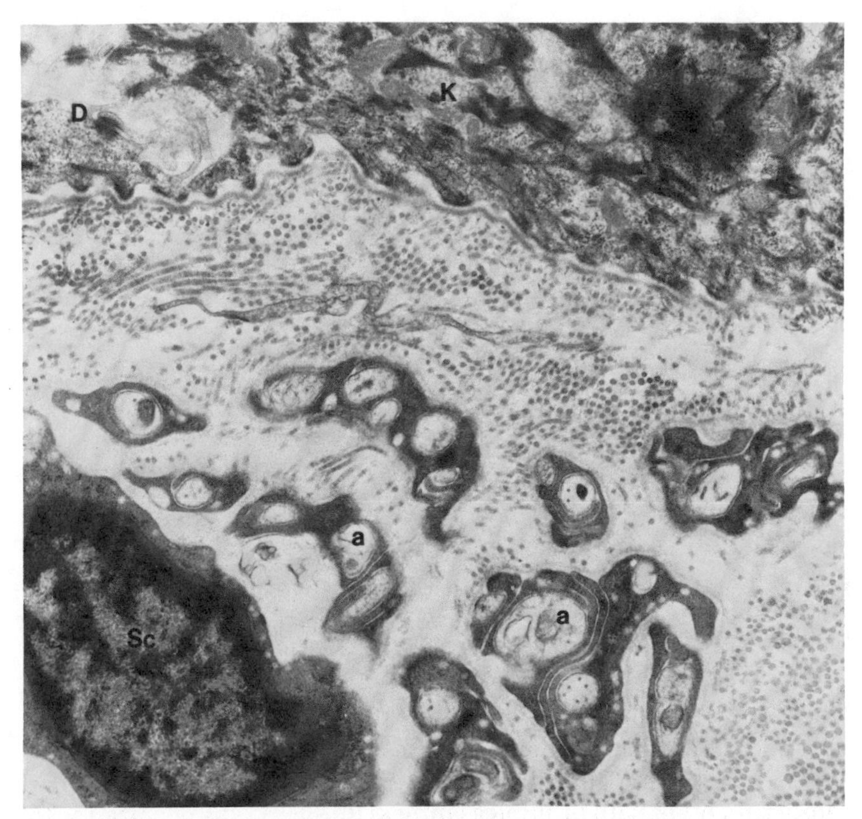

Fig. 11. Electron micrograph of an array of dispersed unmyelinated axons in plexus of dermal fiber and its Schwann cell (*Sc*) underlying basal epidermal keratocytes (*K*) displaying desmosomal junctions (*D*). ×12,400. [From Kruger, Perl, and Sedivec (69).]

whether a fine, isolated axonal profile truly constitutes an end organ. Cauna (22) believed, based on pathological findings, that penicillate terminals might be multimodal receptors, without specifying which modalities might be involved. However, conjectural inferences must ultimately conform with functional data and some morphological sign that distinguishes an ending from an isolated axonal profile. Fine-structural descriptions of cutaneous end organs generally rely on surrounding specializations (sometimes corpuscular) and intra-axonal vesicles and mitochondrial clusters, but most published micrographs of putative free endings usually lack these features. Extensive serial sections might verify a true termination, but this would be an extraordinarily arduous task to perform and document.

The most successful approach to date in functional identification of unmyelinated endings at the fine-structural level derives from examination of the cornea, a structure uniquely innervated by C fibers

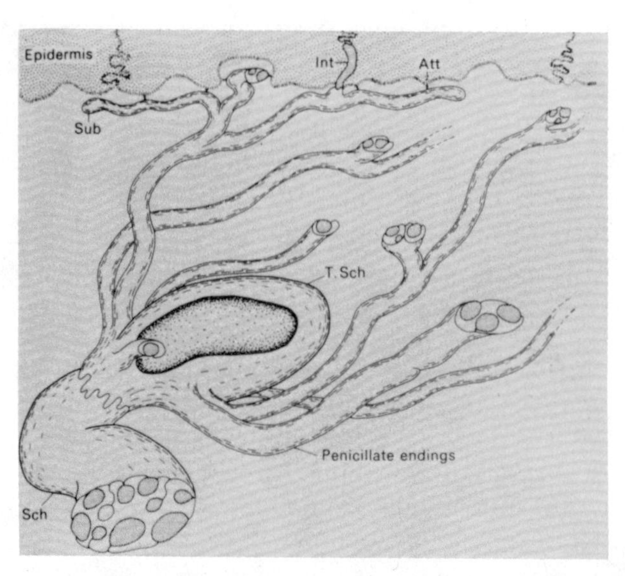

Fig. 12. Diagram of unmyelinated penicillate endings branching from terminal Schwann cell and penetrating epidermis. [From Cauna (21), by permission of Cambridge Univ. Press.]

(107). Although corneal afferent units possess low mechanical thresholds and respond to moderate thermal and chemical stimuli (5, 88, 117), corneal stimulation in humans results in reflex blinking and a sensory report of pain or unpleasantness (6, 90). Axons penetrating the corneal stroma rich in mitochondrial content and possessing a few clear vesicles probably account for the acetylcholinesterase staining of the presumptive endings (53, 112, 113). However, there are numerous terminals containing a mixture of dense-core and clear vesicles, only a few of which degenerate after sympathetic denervation (113). Sinclair's (107) recent monograph expresses reasonable doubt concerning sympathetic innervation of the epidermis, but the presence of some dense-core granular vesicles in axons penetrating the epidermal basal lamina (23, 69) raises the question whether the presence of catecholamines denotes autonomic innervation or whether sensory terminals contain mixtures of agranular and granular vesicles. So far our ignorance concerning the pervasive unmyelinated-fiber population is clearly the principle unresolved problem in cutaneous innervation. Its ultimate resolution can be addressed by advances in recent methodology, principally the intracellular labeling of physiologically identified axons containing a transportable substance and the application of immunocytochemistry for identification of specific receptors of terminal membranes.

Chemical mediation

Chemical mediation of nociceptor excitation and its role in neurogenic inflammatory responses have long been suspected to be a key for understanding many of the peripheral phenomena related to pain. There is some fairly direct evidence of a neurochemical mechanism underlying the phenomenon of nociceptor sensitization (99, 100, 129). The cutaneous flare reaction induced by noxious stimulation, characterized by local vasodilation and plasma extravasation, is believed to be induced by a polypeptide or neurokinin that is clearly neurogenic (i.e., dependent on intact-nerve supply), and infusion of fluid collected in zones of damage can produce vasodilation and pain (25). The neural mechanism of the vascular response is poorly understood, but evidence adduced at the turn of the century by Bayliss for an axon reflex independent of the spinal cord was shown recently to depend solely on thin myelinated and unmyelinated fibers (65). The effect is mimicked by the undecapeptide, substance P. In addition, when small fibers and substance P of primary afferent origin are depleted by neonatal capsaicin treatment, plasma extravasation in response to local application of irritants or antidromic nerve stimulation is abolished (36, 62, 63).

Although a number of other peptides, including somatostatin, neurotensin, vasoactive intestinal peptide, cholecystokinin, angiotensin II, and bombesin, have been identified in sensory nerves (14), the strongest evidence for a peptide association with pain sensation derives from studies of substance P. Its association with cells of sensory ganglia (48) and its peripheral transport to fine fibers is a primary clue. The numerous unmyelinated corneal substance P–immunoreactive fibers are clearly sensory, since they are unaltered by sympathectomy (85). Although the local administration of substance P to the skin does not produce pain directly (14), substance P is known to release histamine, a potent pain-producing substance. Substance P application to spinal neurons produces a powerful excitatory effect on neurons known to be excited by nociceptors (46, 102). Pearson et al. (96) recently showed that in patients with familial dysautonomia with severely impaired pain perception substance P is undetectable in the substantia gelatinosa of the dorsal horn, where it is normally abundant.

A promising approach to elucidating the role of substance P in nociception derives from experiments with capsaicin, the pungent ingredient of paprika and chili peppers, which is a painful skin irritant that subsequently desensitizes nociceptors for long periods (110). Inactivation of small fibers by neonatal capsaicin treatment is especially profound, resulting from degeneration of most of the unmyelinated fibers and perhaps 20–30% of the thinly myelinated population, although there are some conflicting findings on the fiber-spectrum changes (62, 72, 104). Although capsaicin may effect release of several peptides, there is good evidence that it is a fairly specific neurotoxin in some animals (14), which profoundly depletes substance P levels in the dorsal horn of the spinal cord (64).

The behavioral consequences of substance P depletion by capsaicin treatment are not clearly understood, and perhaps factors other than substance P are crucial. A loss of responsiveness to visceral and somatic chemical irritation has been found (17, 44, 62, 111, 128), although there are conflicting findings concerning the effects of substance P depletion on responsiveness to noxious heat (13, 44, 45, 49, 92) and mechanical stimulation (17, 24, 45). There is some suggestion that the profound thermal analgesia reported after intrathecal capsaicin injection (95, 128) may be due to nonspecific cord damage (91). Perhaps the critical factor can be found in the peripheral mechanism, because analgesia results from localized peripheral administration of capsaicin without depleting substance P in ganglion cells (86). In addition, a recently synthesized substance P antagonist, applied locally, dramatically depressed the irritant response to peripheral (cutaneous) application of capsaicin (101). After massive small-fiber depletion produced by neonatal capsaicin treatment, the degree of residual responsiveness to noxious heat and mechanical stimuli suggests that even the preservation of a few small fibers can sustain some aspects of nociceptive function. Such animals may show behavioral withdrawal from noxious heat, although somatovisceral reflexes are abolished (24). The relationship between nociceptive reflexes, sensitivity to noxious stimuli, and pain perception requires some clarification in interpreting such behavioral studies of pain in animals.

Pathology Associated With Nerve Injury

Experiments dealing with the interruption of peripheral nerves or dorsal roots have provided a complex and often controversial literature, especially with respect to pain. In humans, dorsal rhizotomy for the treatment of chronic pain after peripheral nerve injury is often ineffective—a paradox that may now be explained by the recent finding that ganglionectomy does provide the analgesia and relief that should follow complete deafferentation (52). The discovery of unmyelinated afferent fibers in the ventral roots (27), their origin in the dorsal root ganglion (26), and their termination in the pericornual zone of the dorsal horn, where specific nociceptive input is maximal (74), could account for residual pain after dorsal root section.

When a peripheral nerve is injured, a complex series of events is initiated that involves degeneration and regeneration of the nerve distal to the site of damage, the putative release of pain-producing substances from injured nerve terminals and surrounding tissue (73), changes in the excitability and spontaneous discharge of denervated central neurons (77, 79), and reorganization in the viable peripheral (31) and central (33) connections of adjacent intact nerves. Any one of these alterations, peripheral or central, is a potential source of sensory pathology, and in cases of persistent pain after nerve injury, several such changes are probably simultaneously operative.

Peripheral sprouting of nociceptive afferents in noninjured nerves

Even when an injured nerve does not regenerate, the anesthetic zone produced by the lesion gradually shrinks, and recovered cutaneous sensibility is reported to have a protopathic quality (76). It has been suggested that these changes occur because denervated skin is reinnervated with collateral sprouts from high-threshold mechanoreceptive fibers in surrounding intact nerves (31, 60, 93), but evidence for collateral sprouting in adult mammalian skin is equivocal (34), and additional research is necessary to show to what extent sprouting is a factor in postdenervation sensory pathology.

The anatomical evidence for sprouting in skin is based on the inherent limitations of studies that utilized silver-impregnation or methylene blue methods (93, 125, 126). In addition, electrophysiological evidence fails to support the view that the terminal fields of intact nerves in adult animals expand into denervated skin, regardless of whether test stimuli involved the activation of high- or low-threshold mechanoreceptive fibers in the noninjured nerves (31, 50, 61). Devor et al. (31) report that the receptive fields of some small, myelinated units in intact nerves encroach on denervated skin, but in all such examples the apparent invasion of denervated skin was small enough to justify questioning whether the stimulus (pinch) was sufficiently confined (67). The behavioral evidence in animals for the recovery of sensibility to intense mechanical stimulation in denervated skin might be similarly criticized (31, 60, 93). Testing responsiveness to pinch in an awake unrestrained animal is difficult when stimulus control is crude and mechanical spread is uncontrolled. In addition, whenever animals are repeatedly subjected to noxious stimulation in sensate areas, as must be done when spatial alterations in sensate zones is the issue, there is the possibility that behavioral sensitization will confound the results. For example, if a particular cutaneous zone were rendered hypesthetic by nerve injury, pinching the area would not necessarily evoke behavioral signs of sensibility. However, after sensitivity to pinch in both sensate and hypesthetic zones had been repeatedly tested, full-blown responsiveness to any perceived stimulus applied to the hypesthetic zone would probably develop. In this case, sensibility does not recover, but the animal's tendency to respond is altered by repeated noxious stimulation. This interpretation seems likely because behavioral evidence for recovered sensibility after nerve section emerges only when an animal has been either repeatedly tested with noxious pinch stimuli applied to adjacent sensate skin (60) or repeatedly subjected to high-intensity stimulation of adjacent intact nerves (93).

In the final analysis, the only unequivocal evidence for sprouting can be anatomical, and modern histological techniques suitable for visualizing the full fiber complement of nerves in skin have yet to be used for this purpose. The significance of this shortcoming is evinced by our finding that intraspinal sprouting of dorsal root axons is not found when the axons are labeled with a transported tracer (103), in spite of the apparent evidence for such sprouting with silver-impregnation techniques (38, 75).

Neuroma formation

Neuromas are formed in peripheral nerves when injured axons (that fail to grow into the distal nerve stump) form sprouts at the site of damage. Wall et al. (122) have studied the physiological properties of experimentally produced neuromas in animals and have provided evidence supporting the view that afferent activity generated in neuromas contributes to the painful sequelae of peripheral nerve injury. Electrical stimulation of a neuroma produces an afferent barrage in the injured nerve involving primarily small-diameter fibers (32, 40, 122). This probably explains why patients often report that even gentle mechanical stimulation of a neuroma is painful. However, the small-fiber response to electrical stimulation does not preclude the involvement of large fibers when natural stimuli are applied, because it might be more difficult to excite large than small fibers by electrically stimulating the fine terminal sprouts in a neuroma.

Some of the small myelinated afferent fibers that supply a neuroma develop an abnormal spontaneous discharge within 3 days of nerve injury. This aberrant activity peaks within 2 wk, returns to a low level within 1 mo, and remains at the low level for as long as 6 mo (40, 122). Although unmyelinated fibers contribute to the neuroma (32), they have not been isolated for electrophysiological study. The contribution of myelinated fibers to sensory pathology should not be discounted, because selective blockade of A fibers in causalgic patients abolishes the pain associated with innocuous tactile and cold stimulation of hyperalgesic skin, while sensory pathology associated with heat stimulation persists (115).

The sensory axons in a neuroma become sensitive to α-adrenergic agents as their rate of spontaneous discharge increases. Epinephrine or norepinephrine applied locally to a neuroma or administered intravenously (122) and sympathetic trunk stimulation (30) are reported to both increase the firing rate in spontaneously active fibers and activate quiescent fibers. Furthermore the effects of epinephrine can be blocked with α-antagonists (30, 122) but not β-antagonists (66).

Noradrenergic activation of aberrant firing patterns in injured sensory axons is significant in view of the voluminous clinical data implicating the sympathetic nervous system and stress in peripheral neuralgia. Sympathetic trunk stimulation in causalgic patients produces painful sensations referred to the hyperalgesic zone (119), and both guanethidine treatment (43), which blocks the release of norepinephrine from sympathetic terminals, and sympathectomy (9) often provide dramatic relief from causalgic pain. The mechanism(s) for sympathetic activation of somatic afferents is not yet clear. Abnormal sympathetic outflow to hyperalgesic skin (124), ephaptic connections between sympathetic efferent and somatic afferent fibers (30, 105), and aberrant sensory discharge secondary to other effects of sympathetic stimulation, such as peripheral vasoconstriction (66), have been ruled out as contributing factors. Korenman and Devor (66) recently suggested that injured sensory axons develop ectopic α-adrenergic receptors in their terminals within the neuroma and thus become sensitive to α-agents supplied by sympathetic efferent fibers in the damaged nerve and surrounding tissues.

Perhaps the crucial question concerning experimental neuromas is whether the afferent barrages they generate produce pain or dysesthesia. The assessment of spontaneous sensory pathology in animals is difficult, although it has been suggested that self-mutilation in animals is an expression of pain referred to the mutilated body part (78, 121). Sweet (109) recently criticized the validity of this assumption, which is also questioned by a recent observation by one of us (B. Rodin) that the simple manipulation of housing dorsal-rhizotomized males with females prevents the self-mutilation of deafferented hindlimbs that normally occurs when the animals are housed alone. Wall et al. (121) have reported that the incidence of self-mutilation in denervated body parts is correlated with the formation of a neuroma at the site of nerve section, and the time course of the behavior coincides with that of the abnormal discharge in sensory fibers. Treatment with guanethidine, which presumably blocks adrenergic stimulation of neuroma fibers, also prevents self-mutilation (123). If self-mutilation in animals is in fact an expression of sensory pathology, these data may provide the necessary link between the aberrant physiological activity generated in the neuroma and hyperpathia. It should be emphasized, however, that the physiological pathology of the neuroma is short-lived and can only account for sensory pathology in the first few weeks after nerve injury.

The thrust of this discussion has been restricted to attributes of the peripheral nervous system and fails to take central nervous system influences and alterations into account. The complex alterations that must be involved in the profound sensory pathology resulting from nerve and root lesions cannot be accounted for by a simple parsimonious explanation at the peripheral level. A "plastic" reorganization of central connections after peripheral injury may explain many aspects of pain pathology. Whether this is a result of unmasking of inputs to partially denervated cord cells or to sprouting (120) is difficult to resolve, because spatial reorganization is difficult to detect with current anatomical and electrophysiological techniques. However, the conflicting reports on this subject will probably be resolved in this decade.

We are indebted to Anita Roff for preparation and typing of the manuscript.

Our research was supported by National Institutes of Health Grant NS-5685. B. Rodin is a National Institutes of Health Postdoctoral Fellow.

REFERENCES

1. Beck, P. W., and H. O. Handwerker. Bradykinin and serotonin effects on various types of cutaneous nerve fibres. *Pfluegers Arch.* 347: 209–222, 1974.
2. Beck, P. W., H. O. Handwerker, and M. Zimmermann. Nervous outflow from the cat's foot during noxious radiant heat stimulation. *Brain Res.* 67: 373–386, 1974.
3. Bessou, P., P. R. Burgess, E. R. Perl, and C. B. Taylor. Dynamic properties of mechanoreceptors with unmyelinated (C) fibers. *J. Neurophysiol.* 34: 116–131, 1971.
4. Bessou, P., and E. R. Perl. Response of cutaneous sensory units with unmyelinated fibers to noxious stimuli. *J. Neurophysiol.* 32: 1025–1043, 1969.
5. Beuerman, R. W., D. M. Maurice, and D. L. Tanelian. Thermal stimulation of the cornea. In: *Pain in the Trigeminal Region*, edited by D. Anderson and B. Matthews. Amsterdam: Elsevier, 1977, p. 422–423.
6. Beuerman, R. W., and D. L. Tanelian. Corneal pain evoked by thermal stimulation. *Pain* 7: 1–14, 1979.
7. Bishop, G. H., and W. M. Landau. Evidence for a double peripheral pathway for pain. *Science* 128: 712–713, 1958.
8. Blix, M. Experimentelle Beitrage zur Losung der Frage über die specifische Energie der Hautnerven. *Z. Biol. Munich* 20: 141–156, 1884.
9. Bonica, J. J. Causalgia and other reflex sympathetic dystrophies. In: *Advances in Pain Research and Therapy*, edited by J. C. Liebeskind and D. G. Albe-Fessard. New York: Raven, 1979, vol. 3, p. 141–166.

10. Boring, E. G. *Sensation and Perception in the History of Experimental Psychology.* New York: Appleton, 1942.
11. Breathnach, A. S. *An Atlas of the Ultrastructure of Human Skin.* London: Churchill, 1971.
12. Breathnach, A. S. Electron microscopy of cutaneous nerves and receptors. *J. Invest. Dermatol.* 69: 8–26, 1977.
13. Buck, S. H., M. S. Miller, and T. F. Burks. Depletion of primary afferent substance P by capsaicin and dihydrocapsaicin without altered thermal sensitivity in rats. *Brain Res.* 233: 216–220, 1982.
14. Buck, S. H., J. H. Walsh, H. I. Yamamura, and T. F. Burks. Minireview: neuropeptides in sensory neurons. *Life Sci.* 30: 1857–1866, 1982.
15. Burgess, P. R., and E. R. Perl. Myelinated afferent fibres responding specifically to noxious stimulation of the skin. *J. Physiol. London* 190: 541–562, 1967.
16. Burgess, P. R., and E. R. Perl. Cutaneous mechanoreceptors and nociceptors. In: *Handbook of Sensory Physiology. Somatosensory System,* edited by A. Iggo. New York: Springer-Verlag, 1973, vol. II, p. 29–78.
17. Burks, T. F., S. H. Buck, M. S. Miller, P. P. Deshmukh, and H. I. Yamamura. Characterization in guinea pigs of the sensory effects of the putative substance P neurotoxin capsaicin. *Proc. West. Pharmacol. Soc.* 24: 353–357, 1981.
18. Campbell, J. N., R. A. Meyer, and R. H. Lamotte. Sensitization of myelinated nociceptive afferents that innervate monkey hand. *J. Neurophysiol.* 42: 1669–1679, 1979.
19. Campbell, J. N., R. A. Meyer, and S. M. Lancellotta. Correlational analysis of hyperalgesia in humans with responses of nociceptive primary afferents in the monkey. *Soc. Neurosci. Abstr.* 6: 246, 1980.
20. Cauna, N. Light and electron microscopical structure of sensory end-organs in human skin. In: *The Skin Senses,* edited by D. R. Kenshalo. Springfield, IL: Thomas, 1968, p. 15–29.
21. Cauna, N. The free penicillate nerve endings of the human hairy skin. *J. Anat.* 115: 277–288, 1973.
22. Cauna, N. Fine morphological changes in the penicillate nerve endings of human hairy skin during prolonged itching. *Anat. Rec.* 188: 1–11, 1977.
23. Cauna, N. Fine morphological characteristics and microtopography of the free nerve endings of the human digital skin. *Anat. Rec.* 198: 643–656, 1980.
24. Cervero, F., and H. A. McRitchie. Neonatal capsaicin and thermal nociception: a paradox. *Brain Res.* 215: 414–418, 1981.
25. Chapman, L. F., A. O. Ramos, H. Goodell, and H. G. Wolff. Neurohumoral features of afferent fibers in man. Their role in vasodilatation, inflammation, and pain. *Arch. Neurol.* 4: 49–82, 1961.
26. Coggeshall, R. E., M. L. Applebaum, M. Fazen, T. B. Stubbs, and M. T. Sykes. Unmyelinated axons in human ventral roots, a possible explanation for the failure of dorsal rhizotomy to relieve pain. *Brain* 98: 157–166, 1975.
27. Coggeshall, R. E., J. D. Coulter, and W. D. Willis. Unmyelinated axons in the ventral roots of the cat lumbosacral enlargement. *J. Comp. Neurol.* 153: 39–58, 1974.
28. Collins, W. F., F. E. Nulsen, and C. T. Randt. Relation of peripheral nerve fiber size and sensation in man. *Arch. Neurol. Psychiatry* 3: 381–385, 1960.
29. Cunningham, F. O., and M. J. T. Fitzgerald. Encapsulated nerve endings in hairy skin. *J. Anat.* 112: 93–97, 1972.
30. Devor, M., and W. Janig. Activation of myelinated afferents ending in a neuroma by stimulation of the sympathetic supply in the rat. *Neurosci. Lett.* 24: 43–47, 1981.
31. Devor, M., D. Schonfeld, Z. Seltzer, and P. D. Wall. Two modes of cutaneous reinnervation following peripheral nerve injury. *J. Comp. Neurol.* 185: 211–220, 1979.

32. Devor, M., and P. D. Wall. Type of sensory nerve fibre sprouting to form a neuroma. *Nature London* 262: 705–708, 1976.
33. Devor, M., and P. D. Wall. Effect of peripheral nerve injury on receptive fields of cells in the cat spinal cord. *J. Comp. Neurol.* 199: 277–291, 1981.
34. Diamond, J. The recovery of sensory function in skin after peripheral nerve lesions. In: *Posttraumatic Peripheral Nerve Regeneration: Experimental Basis and Clinical Implications*, edited by A. Gorio, H. Millesi, and S. Mingrino. New York: Raven, 1981, p. 533–546.
35. Fitzgerald, M., and B. Lynn. The sensitization of high threshold mechanoreceptors with myelinated axons by repeated heating. *J. Physiol. London* 365: 549–563, 1977.
36. Gamse, R., P. Holzer, and F. Lembeck. Decrease of substance P in primary afferent neurones and impairment of neurogenic plasma extravasation by capsaicin. *Br. J. Pharmacol.* 68: 207–213, 1980.
37. Georgopoulos, A. P. Functional properties of primary afferent units probably related to pain mechanisms in primate glabrous skin. *J. Neurophysiol.* 39: 71–83, 1976.
38. Goldberger, M. E., and M. Murray. Recovery of movement and axonal sprouting may obey some of the same laws. In: *Neuronal Plasticity*, edited by C. W. Cotman. New York: Raven, 1978, p. 73–96.
39. Goldscheider, A. Über den Schmerz. *Gesammelte Abh. Leipzig* 1: 1898.
40. Govrin-Lippmann, R., and M. Devor. Ongoing activity in severed nerves: source and variation with time. *Brain Res.* 159: 406–410, 1978.
41. Halata, Z. The mechanoreceptors of the mammalian skin ultrastructure and morphological classification. *Adv. Anat. Embryol. Cell Biol.* 50: 1–77, 1975.
42. Hallin, R. G., and H. E. Torebjörk. Studies on cutaneous A and C fibre afferents, skin nerve blocks and perception. In: *Sensory Functions of the Skin in Primates*, edited by Y. Zotterman. London: Oxford Univ. Press, 1976, p. 137–148.
43. Hannington-Kiff, J. E. Intravenous regional sympathetic block with guanethidine. *Lancet* 1: 1019–1020, 1974.
44. Hayes, A. G., M. Skingle, and M. B. Tyers. Effects of single doses of capsaicin on nociceptive thresholds in the rodent. *Neuropharmacology* 20: 505–511, 1981.
45. Hayes, A. G., and M. B. Tyers. Effects of capsaicin on nociceptive heat, pressure and chemical thresholds and on substance P levels in the rat. *Brain Res.* 189: 561–564, 1980.
46. Henry, J. L. Effects of substance P on functionally identified units in the cat spinal cord. *Brain Res.* 114: 439–451, 1976.
47. Hensel, H., K. H. Andres, and M. von Düring. Structure and function of cold receptors. *Pfluegers Arch.* 352: 1–10, 1974.
48. Hökfelt, T., J. O. Kellerth, G. Nilsson, and B. Pernow. Substance P: localization in the central nervous system and in some primary sensory neurons. *Science* 190: 889–890, 1975.
49. Holzer, P., R. Jurna, R. Gamse, and F. Lembeck. Nociceptive threshold after neonatal capsaicin treatment. *Eur. J. Pharmacol.* 58: 511–514, 1979.
50. Horch, K. Absence of functional collateral sprouting of mechanoreceptor axons into denervated areas of mammalian skin. *Exp. Neurol.* 74: 313–317, 1981.
51. Horch, K. W., P. R. Burgess, and D. Whitehorn. Ascending collaterals of cutaneous neurons in the fasciculus gracilis of the cat. *Brain Res.* 117: 1–17, 1976.
52. Hosobuchi, Y. The majority of unmyelinated afferent axons in human ventral roots probably conduct pain. *Pain* 8: 167–180, 1980.
53. Hoyes, A. D., and P. Barber. Ultrastructure of corneal receptors. In: *Pain in the Trigeminal Region*, edited by D. Anderson and B. Matthews. Amsterdam: Elsevier, 1977, p. 1–12.
54. Hunt, C. C., and A. K. McIntyre. An analysis of fibre diameter and receptor

characteristics of myelinated cutaneous afferent fibres in cat. *J. Physiol. London* 153: 99–112, 1960.

55. Iggo, A. Cutaneous heat and cold receptors with slowly conducting (C) afferent fibres. *Q. J. Exp. Physiol.* 44: 362–370, 1959.
56. Iggo, A. Cutaneous mechanoreceptors with afferent C fibres. *J. Physiol. London* 152: 337–353, 1960.
57. Iggo, A., and A. R. Muir. The structure and function of a slowly adapting touch corpuscle in hairy skin. *J. Physiol. London* 200: 763–796, 1969.
58. Iggo, A., and H. Ogawa. Primate cutaneous thermal nociceptors (Abstract). *J. Physiol. London* 216: 77P–78P, 1971.
59. Iriuchijima, J., and Y. Zotterman. The specificity of afferent cutaneous C fibres in mammals. *Acta Physiol. Scand.* 49: 267–278, 1960.
60. Jackson, P. C., and J. Diamond. Is sensory nerve activity necessary for collateral sprouting in the skin of adult rats? *Soc. Neurosci. Abstr.* 5: 628, 1979.
61. Jackson, P. C., and J. Diamond. Regenerating axons reclaim sensory targets from collateral nerve sprouts. *Science* 214: 926–928, 1981.
62. Jancsó, G., E. Kiraly, and A. Jancsó-Gábor. Pharmacologically induced selective degeneration of chemosensitive primary sensory neurones. *Nature London* 270: 741–743, 1977.
63. Jancsó, N., A. Jancsó-Gábor, and J. Szolcsányi. Direct evidence for neurogenic inflammation and its prevention by denervation and by pretreatment with capsaicin. *Br. J. Pharmacol.* 31: 138–151, 1967.
64. Jessell, T. M., L. L. Iversen, and A. C. Cuello. Capsaicin induced depletion of substance P from primary sensory neurons. *Brain Res.* 152: 183–188, 1978.
65. Kenins, P. Identification of the unmyelinated sensory nerves which evoke plasma extravasation in response to antidromic stimulation. *Neurosci. Lett.* 25: 137–141, 1981.
66. Korenman, E. M. D., and M. Devor. Ectopic adrenergic sensitivity in damaged peripheral nerve axons in the rat. *Exp. Neurol.* 72: 63–81, 1981.
67. Kruger, L. Cutaneous sense organs and the role of thin fibers in sensation, with particular reference to reinnervation. In: *Posttraumatic Peripheral Nerve Regeneration: Experimental Basis and Clinical Implications*, edited by A. Gorio, H. Millesi, and S. Mingrino. New York: Raven, 1981, p. 549–561.
68. Kruger, L., and S. A. Kroin. A brief historical survey of concepts in pain research. In: *Handbook of Perception*, edited by E. C. Carterette and M. P. Friedman. New York: Academic, 1978, vol. VIB, p. 159–179.
69. Kruger, L., E. R. Perl, and M. J. Sedivec. Fine structure of myelinated mechanical nociceptor endings in cat hairy skin. *J. Comp. Neurol.* 198: 137–154, 1981.
70. Kumazawa, T., and E. R. Perl. Primate cutaneous sensory units with unmyelinated (C) afferent fibers. *J. Neurophysiol.* 40: 1325–1338, 1977.
71. LaMotte, R. H., and J. N. Campbell. Comparison of responses of warm and nociceptive C-fiber afferents in monkey with human judgments of thermal pain. *J. Neurophysiol.* 41: 509–528, 1978.
72. Lasson, S. N., and S. M. Nickels. The use of morphometric techniques to analyse the effect of neonatal capsaicin treatment on rat dorsal root ganglia and dorsal roots (Abstract). *J. Physiol. London* 303: 12P, 1980.
73. Lewis, T. *Pain.* New York: Macmillan, 1942.
74. Light, A. R., and C. B. Metz. The morphology of the spinal cord efferent and afferent neurons contributing to the ventral roots of the cat. *J. Comp. Neurol.* 179: 501–516, 1978.
75. Liu, C. N., and W. W. Chambers. Intraspinal sprouting of dorsal root axons. *Arch. Neurol. Psychiatry* 79: 46–61, 1958.

76. Livingston, W. K. Evidence of active invasion of denervated areas by sensory fibers from neighboring nerves in man. *J. Neurosurg.* 4: 140–145, 1947.
77. Loeser, J. D., and A. A. Ward, Jr. Some effects of deafferentation on the neurons of the cat spinal cord. *Arch. Neurol.* 17: 629–636, 1967.
78. Lombard, M. C., B. S. Nashold, Jr., D. Albe-Fessard, N. Saiman, and C. Sakr. Deafferentation hypersensitivity in the rat after dorsal rhizotomy: a possible animal model of chronic pain. *Pain* 6: 163–174, 1979.
79. Lombard, M. C., B. S. Nashold, Jr., and T. Pelissier. Thalamic recordings in rats with hyperalgesia. In: *Advances in Pain Research and Therapy*, edited by J. C. Liebeskind and D. G. Albe-Fessard. New York: Raven, 1979, vol. 3, p. 767–772.
80. Lynn, B. The heat sensitisation of polymodal nociceptors in the rabbit and its independence of the local blood flow. *J. Physiol. London* 287: 493–507, 1979.
81. Maruhashi, I., K. Mizuguchi, and I. Tasaki. Action currents in single afferent nerve fibres elicited by stimulation of the skin of the toad and the cat. *J. Physiol. London* 117: 129–150, 1952.
82. Melzack, R., and P. D. Wall. On the nature of the cutaneous sensory mechanisms. *Brain* 85: 331–356, 1962.
83. Melzack, R., and P. D. Wall. Pain mechanisms: a new theory. *Science* 150: 971–979, 1965.
84. Meyer, R. A., and J. N. Campbell. Evidence for two distinct classes of unmyelinated nociceptive afferents in monkey. *Brain Res.* 224: 149–152, 1981.
85. Miller, A., M. Costa, J. B. Furness, and I. W. Chubb. Substance P immunoreactive sensory nerves supply the rat iris and cornea. *Neurosci. Lett.* 23: 243–249, 1981.
86. Miller, M. S., S. H. Buck, I. G. Sypes, and T. F. Burks. Capsaicin-induced analgesia: characterization and a site of action. *Soc. Neurosci. Abstr.* 7: 504, 1981.
87. Montagna, W. Morphology of cutaneous sensory receptors. *J. Invest. Dermatol.* 69: 4–7, 1977.
88. Mosso, J. A., and L. Kruger. Spinal trigeminal neurons excited by noxious and thermal stimuli. *Brain Res.* 38: 206–210, 1972.
89. Munger, B. L. Patterns of organization of peripheral sensory receptors. In: *Handbook of Sensory Physiology. Principles of Receptor Physiology*, edited by W. R. Loewenstein. New York: Springer-Verlag, 1971, vol. I, p. 523–556.
90. Nafe, J. P., and D. R. Kenshalo. Somesthetic senses. *Annu. Rev. Psychol.* 13: 201–224, 1962.
91. Nagy, J. I., P. C. Emson, and L. L. Iversen. A re-evaluation of the neurochemical and antinociceptive effects of intrathecal capsaicin in the rat. *Brain Res.* 211: 497–502, 1981.
92. Nagy, J. I., S. R. Vincent, W. A. Staines, H. C. Fibiger, T. D. Reisine, and H. I. Yamamura. Neurotoxic action of capsaicin on spinal substance P neurons. *Brain Res.* 186: 435–444, 1980.
93. Nixon, B., P. Jackson, A. Diamond, A. Foerster, and J. Diamond. Impulse activity evokes collateral sprouting of intact nerves into available target tissue. *Soc. Neurosci. Abstr.* 6: 171, 1980.
94. Orfanos, C. E., and G. Mahrle. Ultrastructure and cytochemistry of human cutaneous nerves. With special reference to the ultrastructural localization of the specific and nonspecific cholinesterases in human skin. *J. Invest. Dermatol.* 61: 108–120, 1973.
95. Palermo, N. N., H. K. Brown, and D. L. Smith. Selective neurotoxic action of capsaicin on glomerular C-type terminals in rat substantia gelatinosa. *Brain Res.* 208: 506–510, 1981.
96. Pearson, J., L. Brandeis, and A. C. Cuello. Depletion of substance P-containing

axons in substantia gelatinosa of patients with diminished pain sensitivity. *Nature London* 295: 61–63, 1982.
97. Perl, E. R. Myelinated afferent fibres innervating the primate skin and their response to noxious stimuli. *J. Physiol. London* 197: 593–615, 1968.
98. Perl, E. R. Is pain a specific sensation? *J. Psychiatr. Res.* 8: 273–287, 1971.
99. Perl, E. R. Sensitization of nociceptors and its relation to sensation. *Adv. Pain Res. Ther.* 1: 17–28, 1976.
100. Perl, E. R. Afferent basis of nociception and pain: evidence from the characteristics of sensory receptors and their projections to the spinal dorsal horn. *Res. Publ. Assoc. Res. Nerv. Ment. Dis.* 58: 19–45, 1980.
101. Piercey, M. F., L. A. Schroeder, K. Folkers, and J.-C. Xu, and J. Horig. Sensory and motor functions of spinal cord substance P. *Science* 214: 1361–1362, 1981.
102. Randic, M., and V. Miletic. Effect of substance P in cat dorsal horn neurons activated by noxious stimuli. *Brain Res.* 128: 164–169, 1977.
102a. Ridley, A. Silver staining of nerve endings in human digital glabrous skin. *J. Anat.* 104: 41–48, 1969.
103. Rodin, B. E., S. Sampogna, and L. Kruger. A reevaluation of intraspinal sprouting of primary afferents. *Soc. Neurosci. Abstr.* 7: 66, 1981.
104. Scadding, J. W. The permanent anatomical effects of neonatal capsaicin on somatosensory nerves. *J. Anat.* 131: 471–484, 1980.
105. Seltzer, Z., and M. Devor. Ephaptic transmission in chronically damaged peripheral nerves. *Neurology* 29: 1061–1064, 1979.
106. Sinclair, D. C. *Cutaneous Sensation.* London: Oxford Univ. Press, 1967.
107. Sinclair, D. C. *Mechanisms of Cutaneous Sensation.* New York: Oxford Univ. Press, 1981.
108. Sinclair, D. C., G. Weddell, and E. Zander. The relationship of cutaneous sensibility to neurohistology in the human pinna. *J. Anat.* 86: 402–411, 1952.
109. Sweet, W. H. Animal models of chronic pain: their possible validation for human experience with posterior rhizotomy and congenital analgesia. *Pain* 10: 275–295, 1981.
110. Szolcsanyi, J. A pharmacological approach to elucidation of the role of different nerve fibers and receptor endings in mediation of pain. *J. Physiol. Paris* 73: 251–259, 1977.
111. Szolcsanyi, J., A. Jancso-Gabor, and F. Joo. Functional and fine structural characteristics of the sensory neuron blocking effect of capsaicin. *Naunyn-Schmiedeberg's Arch. Pharmacol.* 287: 157–169, 1975.
112. Tervo, T. Consecutive demonstration of nerves containing catecholamine and acetylcholinesterase in the rat cornea. *Histochemistry* 50: 291–299, 1977.
113. Tervo, T., F. Joo, K. T. Huikuri, I. Toth, and A. Palkama. Fine structure of sensory nerves in the rat cornea: an experimental nerve degeneration study. *Pain* 6: 57–70, 1979.
114. Torebjörk, H. E. Afferent C units responding to mechanical, thermal and chemical stimuli in human non-glabrous skin. *Acta Physiol. Scand.* 92: 374–390, 1974.
115. Torebjörk, H. E., and R. G. Hallin. Microneurographic studies of peripheral pain mechanisms in man. In: *Advances in Pain Research and Therapy*, edited by J. C. Liebeskind and D. B. Albe-Fessard. New York: Raven, 1979, vol. 3, p. 121–131.
116. Torebjörk, H. E., and J. L. Ochoa. Specific sensations evoked by activity in single identified sensory units in man. *Acta Physiol. Scand.* 110: 445–447, 1980.
117. Tower, S. S. Unit for sensory reception in cornea (with notes on nerve impulses from sclera, iris and lens). *J. Neurophysiol.* 3: 486–500, 1940.
118. Van Hees, J., and J. Gybels. C nociceptor activity in human nerve during painful

and non painful skin stimulation. *J. Neurol. Neurosurg. Psychiatry* 44: 600–607, 1981.

119. Walker, A. E., and F. Nulson. Electrical stimulation of the upper thoracic portion of the sympathetic chain in man. *Arch. Neurol. Psychiatry* 59: 559–560, 1948.
120. Wall, P. D. The role of substantia gelatinosa as a gate control. *Res. Publ. Assoc. Res. Nerv. Ment. Dis.* 58: 205–231, 1980.
121. Wall, P. D., M. Devor, R. Inbal, J. W. Scadding, D. Schonfeld, Z. Seltzer, and M. M. Tomkiewicz. Autonomy following peripheral nerve lesions: experimental anaesthesia dolorosa. *Pain* 7: 103–113, 1979.
122. Wall, P. D., and M. Gutnick. Ongoing activity in peripheral nerves: the physiology and pharmacology of impulses originating from a neuroma. *Exp. Neurol.* 43: 580–593, 1974.
123. Wall, P. D., J. W. Scadding, and M. M. Tomkiewicz. The production and prevention of experimental anesthesia dolorosa. *Pain* 6: 175–182, 1979.
124. Wallin, G., H. E. Torebjörk, and R. G. Hallin. Preliminary observations on the pathophysiology of hyperalgesia in the causalgic pain syndrome. In: *Sensory Function of the Skin in Primates with Special Reference to Man*, edited by Y. Zotterman. Oxford, UK: Pergamon, 1976, p. 489–502.
125. Weddell, G. Axonal regeneration of cutaneous nerve plexuses. *J. Anat.* 77: 49–62, 1942.
126. Weddell, G., L. Guttmann, and E. Gutmann. The local extension of nerve fibres into denervated areas of skin. *J. Neurol. Psychiatry* 4: 206–225, 1941.
127. Willer, N. C., F. Boureau, and D. Albe-Fessard. Role of large diameter cutaneous afferents in transmission of nociceptive messages: electrophysiological study in man. *Brain Res.* 152: 358–364, 1978.
128. Yaksh, T. L., D. H. Farb, S. E. Leeman, and T. M. Jessell. Intrathecal capsaicin depletes substance P in the rat spinal cord and produces prolonged thermal analgesia. *Science* 206: 481–483, 1979.
129. Zimmermann, M. Neurophysiology of nociception. In: *Neurophysiology II*, edited by R. Porter. Baltimore, MD: University Park, 1976, vol. 10, p. 179–221. (Int. Rev. Physiol. Ser.)

TWO

Neurophysiological Mechanisms of Nociception

Ainsley Iggo
Department of Veterinary Physiology,
University of Edinburgh, Edinburgh, Scotland

It can be confidently predicted that neurophysiological mechanisms of nociception in the central nervous system incorporate features consistent with the specialization of the peripheral sensory receptor mechanisms described by Dr. Kruger in chapter one of this book. However, the issue is still debated; in my opinion there is increasing evidence that a specialized pain-signaling mechanism exists within the spinal cord, although there is clearly no simple "private-line" mechanism. Furthermore I strongly maintain that pain is a sensory experience, and thus occurs with certainty in conscious humans with a functional cerebral cortex. The subcortical mechanisms are aspects of nociception—the response of the nervous system (e.g., spinal cord, brain stem, thalamus) to excitation of peripheral nociceptors. Differentiation between nociception and pain aids rational discussion of the relevant mechanisms, in part by clearly delimiting the factors to be considered.

This chapter deals with aspects of the neurophysiological mechanisms of nociception operating at the segmental level of the spinal cord in domestic mammals, principally the cat. Their relevance to sensory pain mechanisms becomes apparent when considering the ascending sensory pathways (discussed in chapt. three).

There are four main aspects at the segmental level: *1*) input, principally along the dorsal roots or trigeminal nerve of the medulla, of impulses coming from various specialized sensory receptors in the peripheral tissues, and the distribution of this input in the spinal cord;

2) dorsal horn neurons and their responses to sensory input, together with the functionally important interactions among these neurons; *3*) influence of descending control mechanisms, originating at several levels of the neuraxis, and how these mechanisms modify dorsal horn neuronal activity (this theme is also discussed in chapts. four to six); and *4*) output from the segmental level in two directions—the ascending sensory pathways (discussed in detail in chapt. three) and output to the motoneurons that leads to muscular activity (i.e., reflex responses). I concentrate on the first two aspects, while preparing the foundation for chapters three to six.

Nociceptive Input and Its Distribution in Spinal Cord

Most nociceptors have fine myelinated or nonmyelinated afferent fibers entering the spinal cord through the dorsal roots (or trigeminal nerve in the case of the medulla). The large myelinated fibers innervate cutaneous mechanoreceptors or muscle and joint mechanoreceptors. Ranson's studies (21) clearly showed that the finer axons, entering via Lissauer's tract, end in the superficial dorsal horn close to their segment of entry. The larger fibers penetrate more deeply into the dorsal horn and in some cases send collateral branches via the dorsal columns to the gracile and cuneate nuclei at the rostral end of the spinal cord.

A striking development in the analysis of the pathways taken by the dorsal root afferent fibers has come from the introduction of horseradish peroxidase (HRP) labeling of neurons and axons, making it possible to observe the ramifications of a single axon or group of axons. Great precision is achieved by injecting HRP into an axon from an indwelling micropipette electrode. In this way the functional characteristics of the afferent fiber (e.g., its sensory receptor, size) can be established by electrical recording. After the injection and appropriate processing, the trajectory of the fiber can be worked out (using computer graphics for expediency). This method was first used for the large myelinated axons from cutaneous and muscle mechanoreceptors and revealed distinctive patterns of terminal arborization of the sensory fibers in the deeper layers of the dorsal horn (2). Technique refinements next revealed the distribution of smaller myelinated axons, including those from cutaneous and muscle nociceptors (22). These results confirmed the expectation that axons from nociceptors had endings in the superficial dorsal horn (in laminae I and II) but also revealed a second, deeper region of termination. We still need detailed single-unit results for the nonmyelinated fibers to reveal the termination fields of identified C mechanoreceptors, thermoreceptors, and nociceptors. Using whole-nerve preparations, Light et al. (15) established that C fibers do terminate in laminae I and II.

This knowledge of spinal cord areas in which different afferent fibers end indicates that neurons in laminae I and II are receiving stations for an input from the nociceptors. Many of the neurons are very small (5–10 μm diam), and morphological studies with light and electron microscopy reveal a complex organization (1, 9, 15, 18).

One new prominent feature is the distribution of neuropeptides in the afferent fibers and neurons of the dorsal horn. This new interest results from several discoveries with immunocytochemical techniques (11, 12). These show that some peptides (e.g., substance P) are present in high concentration in the terminals of nonmyelinated afferent fibers and others (including the opioid peptides) occur in neurons. Capsaicin (an extract from green peppers and capsicum) can deplete substance P from neurons and can prevent the growth and development of nonmyelinated dorsal root fibers in neonatal rats. From growth to maturity, rats exhibit a behavioral indifference to noxious stimuli. Attempts are being made to correlate these two phenomena and to use these special preparations to gain fresh insight into the role of the neuropeptides, particularly in nociception. These new results have sharpened the attack on dorsal horn neurons.

Neurons of Dorsal Horn

Horseradish peroxidase or other intravital cell-labeling techniques have shown that original Golgi-method descriptions of spinal cord neurons were correct in revealing a widespread distribution of the dendrites of individual neurons. Thus a neuron in the middle of the dorsal horn could have dendrites that reach toward the borders of the dorsal horn gray matter in the transverse plane and into adjacent segments in the sagittal plane. Just as with the dorsal root afferent fibers, much greater precision is available with HRP labeling when combined with preliminary electrical recording. The latter process reveals information concerning synaptic inputs to the neuron, its axonal connections, its relative excitability by different kinds of cutaneous afferent input, and its susceptibility to descending control. This is painstakingly slow work, which is gradually developing into a comprehensive picture not only of the neurons that ultimately connect through long ascending axons with higher levels in the sensory pathway but also of any intermediary neurons. One important aspect of this work is the opportunity it provides, through pharmacological study of the neurons and their synaptic connections, to establish highly specific methods for interfering therapeutically with activity in the nociceptive pathways. This prize, I am sure, motivates a good deal of current research.

Classification of dorsal horn neurons

The gray matter of the spinal cord appears stratified or laminated in cross sections, with both the neuronal somata and their dendrites and the axons contributing to the pattern. Based on the cytoarchitectonics, Rexed (23) divided the spinal cord of the cat into 10 laminae (I–X); his system is often used to label neurons (e.g., lamina-V neuron). Some evidence about the functional characteristics initially supported this lamina classification (24), but later results have been more satisfactorily incorporated in a system adopted at an international meeting in Hungary (3, p. 331–332) based on several parameters of the cytoarchitecture and functional characteristics of the neurons. This latest system is also based on the premise that the afferent input comes from functionally specific sensory receptors. Several broad categories are proposed for these neurons, each capable of further subdivision: *1*) mechanoreceptive, excited by an input from mechanoreceptors; *2*) multireceptive, excited by an input from mechanoreceptors and nociceptors; *3*) nocireceptive, excited by an input from nociceptors; and *4*) thermoreceptive, excited by an input from thermoreceptors.

Since the classification is based on experimental results, it is immediately obvious that specialization of the peripheral sensory receptors is matched by a corresponding specialization of the neurons. Because the subject of this book is nociception and pain, I have chosen those neurons that respond to nociceptor inputs (i.e., nocireceptive and multireceptive neurons) as illustrative examples.

Nocireceptive neurons

The superficial laminae of the dorsal horn contain end stations for nociceptive afferent fibers. A typical response of a nocireceptive neuron to noxious stimulation of the skin is illustrated in Figure 1. An important property of these neurons in relation to the neural mechanisms of pain is that they are only excited by an afferent input from peripheral nociceptors, thus suggesting a highly specific sensory pathway for pain. The superficial layers of the dorsal horn are especially rich in neurons of this kind (4–8, 10, 26). A relatively homogeneous group of marginal-zone neurons has been reported in cats, monkeys, and rats. They have little background activity in the absence of noxious stimulation and relatively small receptive fields in the skin, muscles, and/or joints. Some of them, possibly only about one-third (7), send their axons to higher levels of the nervous system, and thus could

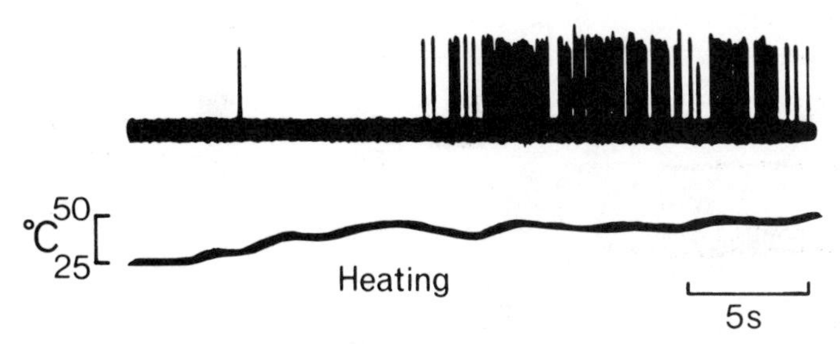

Fig. 1. Response of nocireceptive neuron in lamina I of cat spinal cord to noxious stimulation of skin. [From Cervero, Iggo, and Ogawa (7).]

have an immediate sensory function and be capable of influencing the activity of other neurons in adjacent segments of the spinal cord. There are reports of nocireceptive substantia gelatinosa (SG) neurons, but no detailed description of their properties has yet been made.

Multireceptive neurons

Multireceptive neurons are excited by both an afferent input from mechanoreceptors and nociceptors. They have larger cell bodies than the nocireceptive neurons and were the first neurons studied extensively in the dorsal horn. Studies with selective afferent stimulation, with quantitative control of natural cutaneous and peripheral nerve stimulation (10), have convincingly established that these neurons are excited by various types of cutaneous receptors (Fig. 2). There is much literature on these neurons (for review see ref. 25). The broad group is certainly not homogeneous, either in its peripheral receptive fields, the location of the neuronal somata, or the destination of the axons. Some multireceptive neurons in laminae III and IV project into the spinocervical tract (2), others are postsynaptic neurons that send their axons to dorsal columns, yet others project via the spinoreticular tract to the reticular formation of the brain stem, and some are exclusively segmental in the distribution of their neurons.

Several functional roles have been ascribed to these neurons: T cells in the Melzack-Wall gate hypothesis (16), sensory-discriminative pain signaling by Price and Dubner (20), and diffuse noxious inhibitory control (DNIC) by LeBars et al. (13). Despite a great deal of investigation, the functional roles of these multireceptive neurons are still not clear, and they may, as Lundberg (personal communication) has suggested, be more decisively implicated in reflex activity than in sensation. Undoubtedly they can be powerfully activated by an afferent input from nociceptors and therefore must be seriously considered as puta-

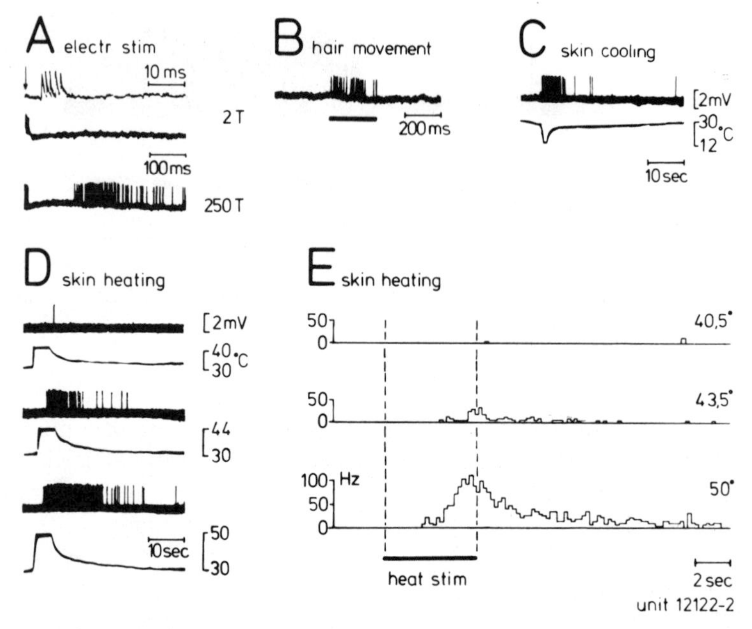

Fig. 2. Multireceptive neuron in laminae IV and V of cat lumbar dorsal horn. Neuron can be excited by a variety of natural stimuli and by electrical stimulation of both large and small afferent fibers. [From Handwerker, Iggo, and Zimmermann (10).]

tive pain neurons, although more exclusively nocireceptive neurons exist.

Substantia gelatinosa

The SG lying between the larger neurons of laminae I and III receives a rich afferent input from small afferent fibers (see **Nociceptive Input and Its Distribution in Spinal Cord**). The older histological descriptions with light microscopy and silver stains have now been dramatically extended with the new techniques of neuronal labeling with HRP and of immunocytochemistry (3). It is significant for nociception that most nonmyelinated afferent fibers have end stations in the SG. The SG probably has a role in nociception, because a varying but significant proportion in different species of the nonmyelinated dorsal root axons is nocireceptive (for review see ref. 5).

The functional characteristics of identified SG neurons have only recently been accessible to electrophysiological investigation. Sustained intracellular examination of these very small neurons is now possible. Labeling them with HRP verifies the locations of their cell bodies as well as the distribution of their dendrites and axons (1, 15, 18). As might be expected, several kinds of response patterns have

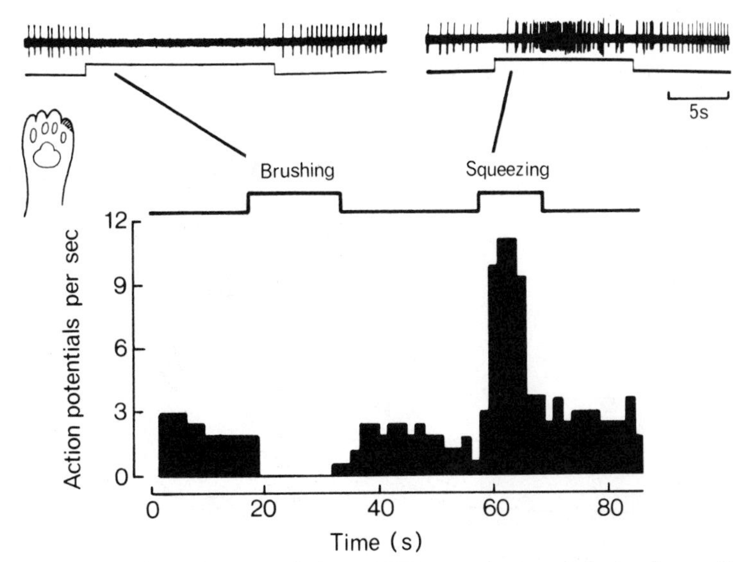

Fig. 3. Responses of substantia gelatinosa (SG) neuron to natural stimulation. Ongoing regular background discharge was strongly inhibited by lightly brushing hairs on toe and enhanced by squeezing same region of skin. This is an inverse 1 ($\bar{1}$) SG unit. (Courtesy of V. Molony and W. M. Steedman.)

emerged, suggesting a variety of functions for these neurons. In my laboratory, particular attention has been paid to neurons with a persistent background discharge, which may continue uninterrupted for many minutes, in the absence of any intentionally applied sensory stimulation. This sustained activity can be reduced or abolished in a graded manner by appropriate sensory stimulation (Fig. 3), and the neurons can be grouped into classes corresponding to those already enumerated for dorsal horn neurons in general. Note, however, that the actions are the reciprocal of those previously described, e.g., nocireceptive SG neurons that are inhibited rather than excited by noxious stimulation. These results led us to classify these neurons as inverse (6) and to suggest a role for them in the reciprocal regulation of dorsal horn neurons in laminae I, IV, and V (4).

It is now possible to propose a simplified scheme for the neuronal organization of the dorsal horn of the spinal cord based on results obtained by new methods (Fig. 4), such as those mentioned. A similar functional scheme is more difficult to prepare.

Segmental Interactions

The characteristics of the dorsal horn have so far been principally considered in the light of excitatory neuronal responses to an afferent

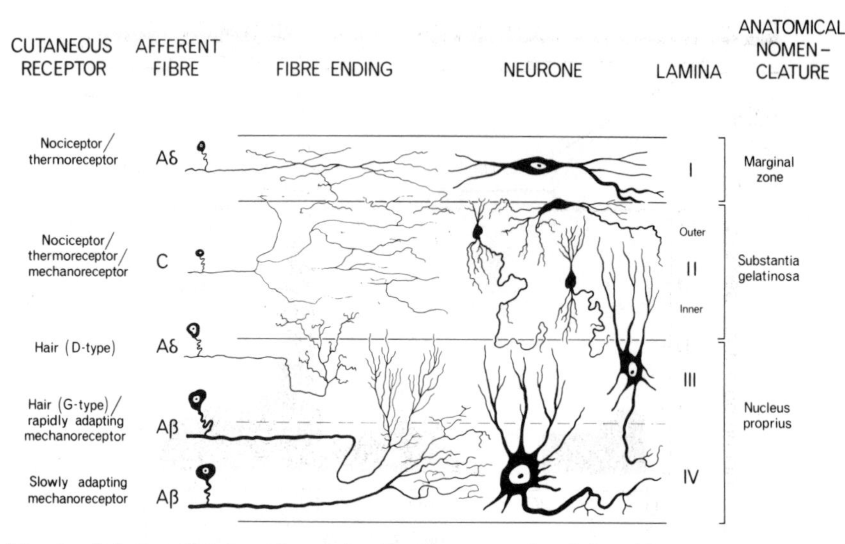

Fig. 4. Schema of neuronal organization of cat superficial dorsal horn. [From Cervero and Iggo (5).]

input from the periphery. I now turn to another important property of the neurons—the ability to be inhibited as well as excited. The basic mechanisms are synaptic actions, either excitatory caused by depolarization or inhibitory. The latter may be *1*) postsynaptic, in which there is hyperpolarization of the neuron, or *2*) presynaptic, by reduction in the efficacy of excitatory axons.

The action of noxious afferent inputs on nocireceptive and multireceptive neurons can be locally inhibited. Figure 5 illustrates this effect for both kinds of neuron. The experiments established that inhibition could be initiated by impulses from sensitive cutaneous mechanoreceptors, although not from the large muscle afferent systems. Figure 5 illustrates that inhibition had a short latency of onset and did not long outlast the inhibitory stimulus. It could also be graded in intensity. Mechanisms therefore exist in the spinal cord that result in an interplay of excitatory and inhibitory stimuli arising from various kinds of specific sensory receptors. Thus mechanoreceptors can inhibit the actions of nociceptors, an effect that may partly account for the relief of itch or soreness by scratching or rubbing the skin. Of course scratching may also work by removing the source of the irritation.

The detailed neuronal mechanisms by which this inhibitory-excitatory interplay works are not yet known, but they certainly involve local action by the neurons already described.

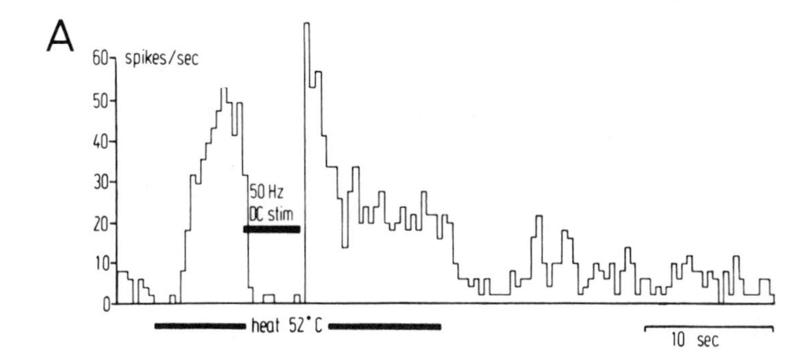

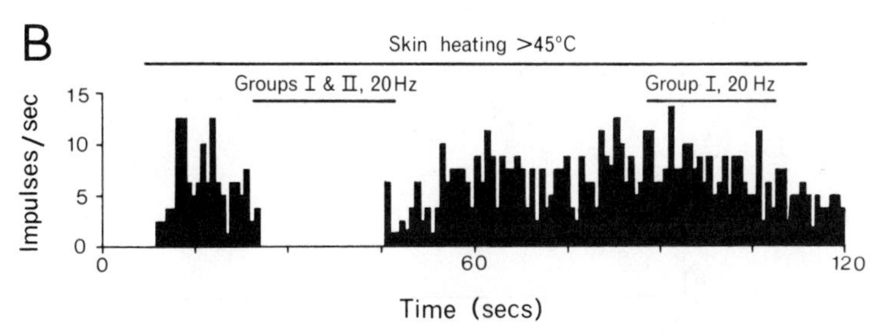

Fig. 5. Segmental inhibition of multireceptive neuron (*A*) and nocireceptive neuron (*B*). *A*: multireceptive neuron was first excited by heating skin to 52°C, and while stimulus continued, large afferent fibers in dorsal columns (DC) were stimulated electrically. This caused an immediate reduction in response to skin heating. Latter response recovered rapidly when dorsal column stimulation was removed. [From Handwerker, Iggo, and Zimmermann (10).] *B*: discharge in nocireceptive neuron caused by heating skin to >45°C was inhibited by electrical stimulation of large (group I and II) afferent fibers in the tibial nerve, although not by activation of only group I (i.e., large muscle afferent fibers). [From Cervero, Iggo, and Ogawa (7).]

Descending Control Mechanisms

In addition to local interactions, the brain can influence the sensitivity of neurons in the dorsal horn by descending influences arising in the brain stem or more cranially. An example is illustrated in Figure 6, which shows the effect of spinal block on the responses of a multireceptive neuron. A noxious stimulus to the footpad could not excite the neuron when the spinal cord was intact, whereas it caused a vigorous discharge when the spinal cord was blocked. This is evidence for a powerful tonic descending inhibitory action. Of even greater interest is the selective nature of the action. The response of the same neuron to excitatory tactile stimuli was almost the same in both intact and blocked spinal cords. A reasonable explanation is that

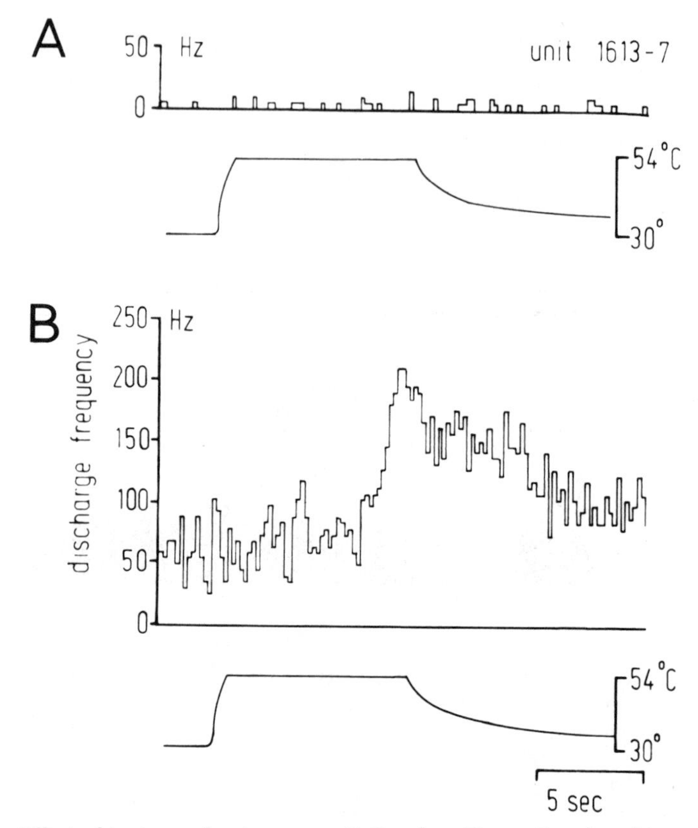

Fig. 6. Effect of tonic mechanisms on activity of multireceptive dorsal horn neuron. *A*: lack of response of neuron when skin was heated, with spinal cord intact; *B*: powerful response when spinal cord was blocked, thus cutting inhibitory mechanisms. [From Handwerker, Iggo, and Zimmermann (10).]

the descending action was expressed on an interneuron in the excitatory pathway from the nociceptor to the multireceptive neuron in lamina IV. The interneurons involved have not been identified, but the possibility that they are SG neurons is ruled out by the almost complete absence of change in the properties of SG neurons studied in the intact or blocked spinal cord (6). However, the excitability of the neurons could be altered by electrical stimulation of the ipsilateral cord rostral to the recorded SG neurons. The nocireceptive neurons are also under tonic control, but this appears weaker than that expressed for multireceptive neurons (7).

Phasic Descending Control

Considerable interest has been aroused by reports of behavioral

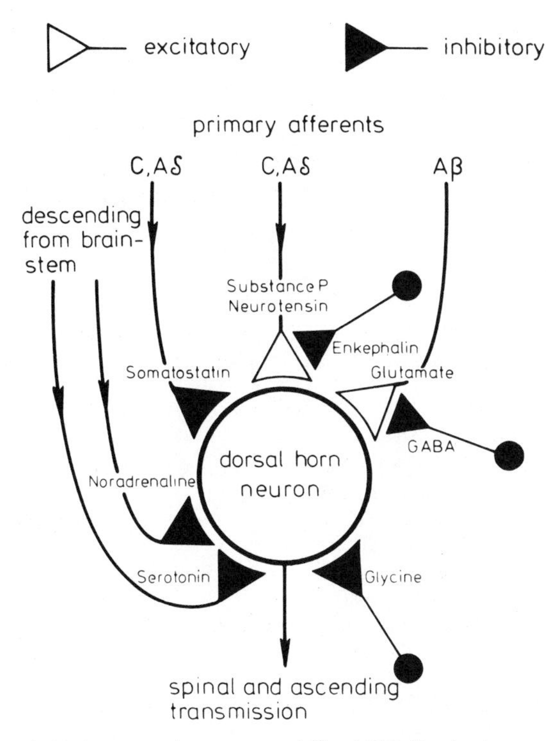

Fig. 7. Schematic illustration of some synaptic mechanisms and transmitters known or postulated to modify activity of multireceptive and nocireceptive dorsal horn neurons. [From M. Zimmermann (27), *Advances in Pain Research and Therapy*. New York: Raven, 1979.]

experiments, establishing that the sensitivity of an animal to noxious stimulation can be dramatically reduced by electrical stimulation of discrete regions of the brain stem (14). Neurophysiological studies suggest that an endogenous analgesic system exists, which at least partly operates on dorsal horn neurons via tryptaminergic pathways. The endogenous opioids have been implicated at brain stem and/or spinal levels. Morphine, the morphine antagonist naloxone, and endogenous opioids all contribute to the elucidation of the mechanisms involved. In addition to these chemicals there is good evidence that descending inhibition can be mediated via axons containing serotonin that have cell bodies in the nucleus raphe magnus of the brain stem (19). The nucleus locus ceruleus containing norepinephrine also has an inhibitory action on dorsal horn neurons (17).

The results briefly outlined here indicate that the powerful control exerted over the spinal cord by the higher centers in the brain are multivariate, which suggests the possibility of many ways for activating endogenous analgesic mechanisms. Concomitant developments in neurochemistry also prognosticate important developments in the neuropharmacology of the control of analgesia. Figure 7 shows the numerous ways in which dorsal horn neurons can have their activity modulated.

REFERENCES

1. Bennett, G. J., M. Abdelmoumene, H. Hayashi, and R. Dubner. Physiology and morphology of substantia gelatinosa neurons intracellularly stained with horseradish peroxidase. *J. Comp. Neurol.* 194: 809–827, 1980.
2. Brown, A. G. Spinocervical tract neurones. In: *Organization in the Spinal Cord.* Berlin: Springer-Verlag, 1981, p. 73–104.
3. Brown, A. G., and M. Réthelyi. Reports of working parties. In: *Spinal Cord Sensation.* Edinburgh: Scottish Academic, 1981, p. 331–336.
4. Cervero, F., and A. Iggo. Reciprocal sensory interaction in the spinal cord (Abstract). *J. Physiol. London* 284: 84P–85P, 1978.
5. Cervero, F., and A. Iggo. The substantia gelatinosa of the spinal cord: a critical review. *Brain* 103: 717–772, 1980.
6. Cervero, F., A. Iggo, and V. Molony. Segmental and intersegmental organization of neurones in the substantia gelatinosa Rolandi of the cat's spinal cord. *Q. J. Exp. Physiol.* 64: 315–325, 1979.
7. Cervero, F., A. Iggo, and H. Ogawa. Nociceptor-driven dorsal horn neurones in the lumbar spinal cord of the cat. *Pain* 2: 5–24, 1976.
8. Christensen, B. N., and E. R. Perl. Spinal neurons specifically excited by noxious or thermal stimuli: marginal zone of the dorsal horn. *J. Neurophysiol.* 33: 293–307, 1970.
9. Gobel, S., W. M. Falls, G. J. Bennett, M. Abdelmoumene, H. Hayashi, and E. Humphrey. An EM analysis of the synaptic connections of horseradish peroxidase-filled stalked cells and islet cells in the substantia gelatinosa of adult cat spinal cord. *J. Comp. Neurol.* 194: 781–807, 1980.
10. Handwerker, H. O., A. Iggo, and M. Zimmermann. Segmental and supraspinal actions on dorsal horn neurons responding to noxious and non-noxious skin stimuli. *Pain* 1: 147–165, 1975.
11. Hökfelt, T., O. Johansson, A. Ljungdahl, J. M. Lundberg, and M. Schultzberg. Peptidergic neurones. *Nature London* 284: 515–521, 1980.
12. Hunt, S. P., P. C. Emson, R. Gilberg, M. Goldstein, and J. R. Kimmell. Presence of avian pancreatic polypeptide-like immunoreactivity in catecholamine and methionine-enkephalin-containing neurons within the central nervous system. *Neurosci. Lett.* 21: 125–130, 1981.
13. LeBars, D., A. H. Dickenson, and J.-M. Besson. Diffuse noxious inhibitory controls (DNIC). I. Effects on dorsal horn convergent neurons in the rat. *Pain* 6: 283–304, 1979.
14. Liebeskind, J. C., G. Guilbaud, J.-M. Besson, and J. L. Oliveras. Analgesia from electrical stimulation of the periaqueductal gray matter in the cat. *Brain Res.* 50: 441–446, 1973.
15. Light, A. R., D. L. Trevino, and E. R. Perl. Morphological features of functionally defined neurons in the marginal zone of substantia gelatinosa of the spinal dorsal horn. *J. Comp. Neurol.* 186: 151–171, 1979.

16. Melzack, R., and P. D. Wall. Pain mechanisms: a new theory. *Science* 150: 971–979, 1965.
17. Mokha, S. S., J. A. McMillan, and A. Iggo. Descending influences on spinal nociceptive neurones from locus coeruleus: actions, pathways, neurotransmitters and mechanisms. In: *Advances in Pain Research and Therapy*, edited by J. J. Bonica, U. Lindblom, and A. Iggo. New York: Raven, 1983, vol. 5, p. 387–392.
18. Molony, V., W. M. Steedman, F. Cervero, and A. Iggo. Intracellular marking of identified neurones in the superficial dorsal horn of the cat spinal cord. *Q. J. Exp. Physiol.* 66: 211–223, 1981.
19. Oliveras, J. L., G. Guilbaud, and J.-M. Besson. A map of serotoninergic structures involved in stimulation producing analgesia in unrestrained freely moving cats. *Brain Res.* 164: 317–322, 1979.
20. Price, D. D., and R. Dubner. Neurons that subserve the sensory-discriminative aspects of pain. *Pain* 3: 307–338, 1977.
21. Ranson, S. W. Unmyelinated nerve-fibers as conductors of protopathic sensation. *Brain* 38: 381–389, 1915.
22. Réthelyi, M., A. R. Light, and E. R. Perl. Synaptic complexes formed by functionally defined primary afferent units with fine myelinated fibers. *J. Comp. Neurol.* 207: 381–393, 1982.
23. Rexed, B. The cytoarchitectonic organization of the spinal cord in the cat. *J. Comp. Neurol.* 96: 415–495, 1952.
24. Wall, P. D. Dorsal horn electrophysiology. In: *Handbook of Sensory Physiology. Somatosensory System*, edited by A. Iggo. Berlin: Springer-Verlag, 1973, vol. II, p. 253–270.
25. Willis, W. D., and R. E. Coggeshall. Functional organisation of dorsal horn interneurons. In: *Sensory Mechanisms of the Spinal Cord*. New York: Wiley, 1978, p. 129–166.
26. Willis, W. D., D. L. Trevino, J. D. Coulter, and R. A. Maunz. Responses of primate spinothalamic tract neurons to natural stimulation of hindlimb. *J. Neurophysiol.* 37: 357–372, 1974.
27. Zimmermann, M. Peripheral and central nervous mechanisms of nociception, pain and pain therapy. In: *Advances in Pain Research and Therapy*, edited by J. J. Bonica, J. C. Liebeskind, and D. G. Albe-Fessard. New York: Raven, 1979, vol. 3, p. 3–32.

THREE

Ascending Pathways Transmitting Nociceptive Information in Animals

William D. Willis, Jr.

Marine Biomedical Institute, Department of Physiology and Biophysics and Department of Anatomy, University of Texas Medical Branch, Galveston, Texas

The responses of animals to noxious stimuli are comparable to those of humans and thus include both sensory and motor reactions (37). Furthermore the sensory component of the pain response is divisible into sensory-discriminative and motivational-affective aspects (61). This chapter emphasizes the ascending tracts of the spinal cord that are likely to play an important role in the sensory responses to noxious stimuli in animals, especially in common laboratory animals such as the monkey, cat, and rat.

First, the evidence for the parts of the spinal cord white matter that contain ascending tracts important for nociceptive reactions (other than reflexes) is discussed. Second, anatomical evidence for the specific ascending tracts present in these parts of the spinal cord white matter is considered. Finally, the response properties of identified neurons giving rise to the tracts found in the spinal white matter in areas required for nociception are evaluated for the capacity of these cells to transmit nociceptive information to the brain suitable for triggering sensory-discriminative or motivational-affective processes. Some of the evidence included here has been reviewed previously (78), but other findings are more recent.

Location of Nociceptive Tracts in Spinal White Matter

Most of the evidence concerning the location of nociceptive tracts in the spinal cord white matter comes from clinical studies of humans in whom lesions, either due to disease or deliberate surgical intervention, have interrupted pathways in the anterolateral part of the cord, with a resultant contralateral analgesia (75, 78). There have been a few comparable studies in animals, however, where surgical lesions were placed in various parts of the white matter to determine the regions needed for the animal to respond appropriately to noxious stimulation.

The arrangement of the ascending nociceptive pathways in the monkey seems most similar to that of humans. For instance, Yoss (81) found that interruption of the ventrolateral quadrant of the monkey spinal cord results in a loss of responsiveness to noxious compression of the Achilles tendon of the contralateral hindlimb. Vierck and Luck (74) trained monkeys to escape electrical shocks applied to the legs by pressing a panel with a force high enough to discourage responses to innocuous or to minimally noxious stimulus intensities. A lesion of the ventral part of the lateral funiculus produced a temporary analgesia of the contralateral hindlimb. However, the analgesia persisted longer if the entire ventrolateral quadrant was interrupted (i.e., including the ventral funiculus). The most enduring analgesia resulted from a lesion that transected the ventral half of the spinal cord. Lesions of the dorsal column plus Lissauer's tract also attenuated nociceptive responses in monkeys, but lesions of the dorsal part of the lateral funiculus made the animals hyperalgesic (73). Evidently the most important nociceptive pathways in the monkey are in the white matter of the ventral half of the spinal cord.

The situation in the cat seems to be quite different from that in the monkey. Kennard (43) found that cats subjected to an intrathecal injection of alumina cream became hyperalgesic. A hemisection or interruption of both ventral quadrants of the spinal cord produced no apparent reduction in the hyperalgesia, but a lesion of the dorsal parts of both lateral funiculi was effective. Some nociceptive responsiveness remained, however, indicating that nociceptive pathways are present in dorsal and ventral locations in the cat spinal cord.

Casey et al. (16) recently confirmed Kennard's findings. The probability of a response by a cat to a noxious heat stimulus applied to the thigh was reduced by lesions of the thoracic spinal cord interrupting either the ventral white matter or the dorsal columns and dorsal lateral funiculi. However, the response probability was increased by a lesion interrupting the deeper parts of the dorsal white matter. It was concluded that in cats there are ascending pathways important for pain in both the dorsal and ventral parts of the cord and pain-suppressing paths between the dorsally and ventrally situated pain pathways. A dorsal column lesion had only a transient effect.

The rat seems to be different still. Basbaum (5) was unable to eliminate an escape response in rats even after two staggered hemisections on opposite sides of the spinal cord. It was concluded that the ascending nociceptive pathways in the rat include a propriospinal component that allows transmission of nociceptive information up the cord despite interruption of the long ascending tracts. However, this does not rule out an important contribution of the long tracts to nociception in the rat. Similar observations were made in the pig by Breazile and Kitchell (12).

These species differences in the effects of spinal cord lesions on nociceptive responses can be expected to be reflected in differences in the organization and function of nociceptive neurons in different animals.

Somatosensory Tracts Ascending in Ventrolateral White Matter

The somatosensory pathways known to project in the ventrolateral white matter of the spinal cord include the spinothalamic, spinomesencephalic, and spinoreticular tracts. These pathways are found in the monkey (6, 8, 10, 46, 54), cat (6, 7, 41, 60), and rat (28, 30, 51, 53), as well as in many other vertebrate species (11, 23, 38, 42, 53, 63).

The spinothalamic tract is largely a crossed pathway (Fig. 1; 15, 28, 79). It terminates in several different nuclei of the thalamus, including the ventral posterior lateral nucleus in the monkey and rat or a shell region immediately surrounding this nucleus in the cat (6–8, 41, 51, 54). There are also terminals in the intralaminar complex, especially the central lateral nucleus, and in certain other nuclei of the thalamus [e.g., the medial part of posterior complex and the nucleus submedius (21)].

The spinomesencephalic pathway can probably be considered in part an extension of the spinoreticular system (Fig. 1). Although there are important connections with the superior colliculus (3), the ascending spinal projections to such midbrain nuclei as the periaqueductal gray and the mesencephalic reticular formation are most relevant here (11, 38, 46, 47, 54, 63). The spinomesencephalic tract in the monkey, like the spinothalamic tract, is largely crossed (69, 79). In the rat, however, studies with antidromic activation indicate large numbers of uncrossed connections to the midbrain (55). The axons of some of these cells may cross to the contralateral side within the midbrain (3).

The spinoreticular tract projects both to the lateral reticular nucleus

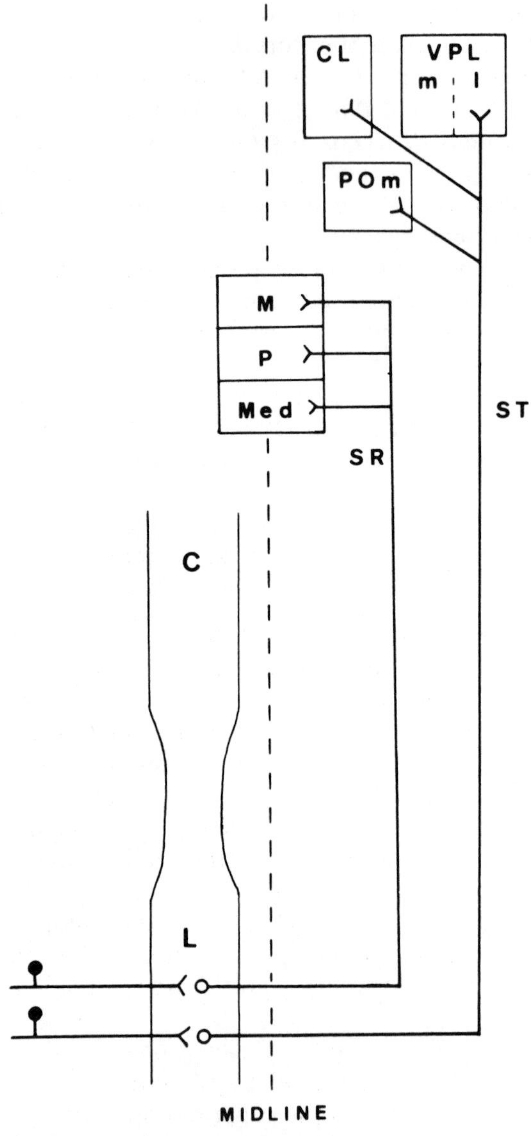

Fig. 1. Organization of candidate nociceptive pathways in ventrolateral white matter of spinal cord. Spinothalamic (ST) and spinoreticular (SR) tracts from lumbosacral (L) cord are shown. Comparable pathways also originate from other levels of spinal cord, including cervical enlargement (C). However, spinoreticular projections from cervical enlargement are bilateral. Nuclei of termination of ST or SR tracts include medial and lateral parts of ventral posterior lateral (VPL_m, VPL_l), central (CL) and medial part of posterior (PO_m) complex of thalamus, and midbrain, pontine, and medullary reticular formation (M, P, and Med). [From Willis and Coggeshall (78).]

[a cerebellar relay (19)] and to the medial reticular formation (10, 11, 46, 54). There are terminations in several medial reticular nuclei, including the nucleus centralis medulla oblongata, the nucleus reticularis gigantocellularis, the nucleus reticularis pontis caudalis and oralis, and the nucleus paragigantocellularis dorsalis and lateralis. There are also projections to the nucleus subceruleus and perhaps to the raphe nuclei. Whether the spinoreticular tract is primarily a crossed or an uncrossed pathway is controversial. Kerr and Lippman (47) found very little terminal degeneration in the reticular formation (except in the caudal medulla) after multisegmental commissural myelotomies in the monkey, whereas the same complement of degeneration was observed in the thalamus as is seen after anterolateral cordotomy. These authors suggest that the spinoreticular tract must be largely uncrossed, in contrast to the spinothalamic tract. However, a recent horseradish peroxidase (HRP) study by Kevetter et al. (48) in the monkey indicates that the spinoreticular tract originating from the cervical enlargement is bilateral, whereas that from the lumbosacral enlargement is largely crossed. Note that all successful commissural myelotomies in the study by Kerr and Lippman (47) were done at the level of the cervical enlargement. Thus the results of these two studies are not mutually incompatible.

Since the HRP technique for retrograde labeling became available, the locations of the cells of origin of the ascending pathways from the spinal cord have received much attention. The cells of origin of the spinothalamic tract in the monkey are widely distributed in the spinal cord gray matter (70, 79). However, certain regions of concentration can be seen after focal injections of HRP into the region of the ventral posterior lateral (VPL) nucleus or into the region of the central lateral nucleus (Fig. 2*A*; 79). Cells projecting to the VPL nucleus region are found chiefly in layers of the dorsal horn equivalent in the monkey to laminae I and V of Rexed (64) in the cat. Injections of HRP into the intralaminar region label cells concentrated in laminae VI–VIII. However, there is substantial overlap in the distribution of spinothalamic cells projecting laterally versus medially in the thalamus of the monkey (79).

The location of the cells of origin of the spinothalamic tract in the cat cervical enlargement is similar to that described for the monkey. However, spinothalamic tract cells in the cat lumbosacral spinal cord are rarely found in lamina V; instead they are most prevalent in laminae I, VII, and VIII (Fig. 2*B*; 15, 71). Cells projecting to the intralaminar region are mostly in laminae VII and VIII in both the cervical and lumbosacral enlargements (Fig. 2*B*; 15).

The spinothalamic tract of the rat originates from cells in both the

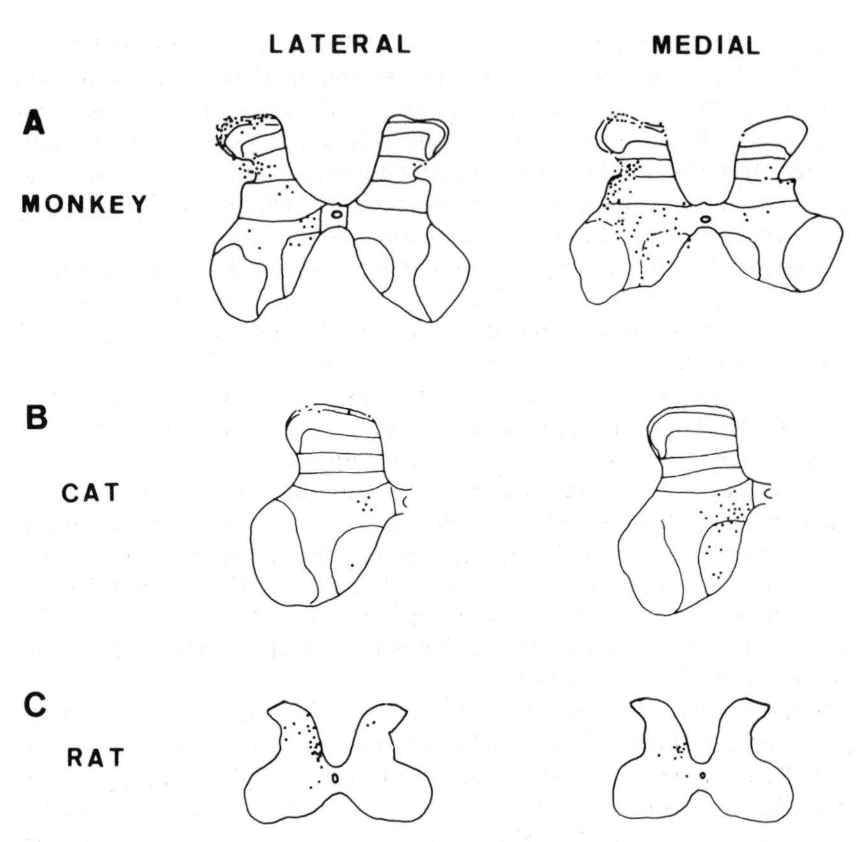

Fig. 2. Locations of cells of origin of spinothalamic tract. Spinothalamic cells were labeled by retrograde transport of HRP injected into either lateral or medial thalamus. *A*: monkey [from Willis et al. (79)]; *B*: cat [from Carstens and Trevino (15)]; and *C*: rat [from Giesler et al. (28)].

dorsal and ventral horns, like in the monkey and cat (Fig. 2*C*; 28, 30). However, only a few spinothalamic cells are in lamina I. The cells projecting to the VPL nucleus are commonly found superficially in the dorsal horn, although many are in the medial base of the dorsal horn and a few are in the ventral horn. Those projecting to the intralaminar complex are most often located in the base of the dorsal horn and in the intermediate zone (28, 30).

Spinomesencephalic cells in the monkey and rat are located in the same area as that portion of the spinothalamic cell population projecting to the VPL nucleus (Fig. 3*A*; 55, 69, 79). Electrophysiological evidence suggests that many of the same neurons in the monkey send axonal branches both to the VPL nucleus and to the midbrain (62).

The cells of origin of the spinoreticular projection to the medulla,

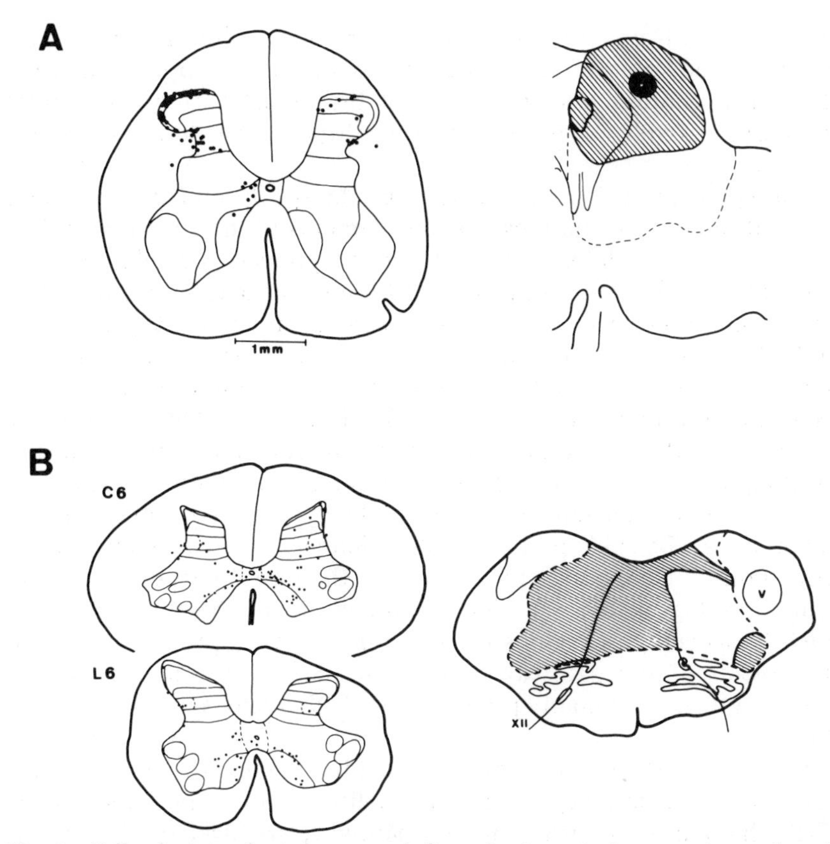

Fig. 3. Cells of origin of spinomesencephalic and spinoreticular tracts in monkey. *A*: distribution of cells labeled by retrograde transport of HRP from midbrain shown in lumbosacral enlargement of monkey (*left*). Most labeled cells are contralateral to injection site (*right*). [From Willis et al. (79).] *B*: distribution of cells labeled by HRP injected bilaterally into pontomedullary reticular formation shown for segments C_6 and L_6 (*left*). Injection was centered in medulla (*right*). [From Kevetter et al. (48).]

on the other hand, are chiefly in the same locations as cells projecting to the intralaminar complex of the thalamus, at least in the monkey and rat (Fig. 3*B*; 48, 49). In fact many individual spinal neurons in the rat project both to the medullary reticular formation and to the thalamus (49).

Somatosensory Tracts Ascending in Dorsal Part of Spinal Cord

At least two pathways have been identified that ascend in the dorsal part of the spinal cord and may convey nociceptive information: the spinocervical tract and the second-order dorsal column pathway. Included in the latter, for present purposes, are cells whose axons reach the dorsal column nuclei by way of the dorsal part of the lateral funiculus as well as those projecting through the dorsal funiculus proper.

The spinocervical tract ascends ipsilaterally to end in the lateral cervical nucleus in the most rostral segments of the spinal cord (Fig. 4; 20, 59, 65). The lateral cervical nucleus projects contralaterally by way of the medial lemniscus to the VPL nucleus of the thalamus (6, 9, 34). A lateral cervical nucleus is present in monkeys (58), cats (65), and other mammals, like the raccoon (35), but it was initially difficult to recognize in the rat and other species having a lateral spinal nucleus extending the length of the cord (33). Giesler et al. (28, 31) recently showed that the rostralmost part of the lateral spinal nucleus of the rat is equivalent to the lateral cervical nucleus of other mammals.

The second-order dorsal column pathway was first noticed in the electrophysiological experiments of Uddenberg (72). The axons ascend ipsilaterally in the dorsal funiculus and the dorsal part of the lateral funiculus (Fig. 4; 66, 67). The terminals are in the dorsal column nuclei, especially in regions surrounding the main projection area of the dorsal column pathway proper. The cells of the dorsal column nuclei in turn presumably project to the VPL nucleus of the thalamus (Fig. 4).

The cells of origin of the spinocervical tract and of the second-order dorsal column pathway have been delimited by experiments with HRP. The neurons of origin of both pathways are located primarily in lamina IV (Fig. 5; 20, 68).

Response Properties of Neurons in Candidate Nociceptive Pathways

Although behavioral and anatomical evidence can suggest which pathways may be important for nociception in various animals, electrophysiological evidence that the response properties of cells of the candidate nociceptive pathways are consistent with a role in nociception is necessary. The following sections review the response patterns of neurons belonging to the spinothalamic, spinoreticular, spinocervical, and second-order dorsal column pathways.

Spinothalamic tract

Most neurons of the spinothalamic tract (STT) so far investigated in the monkey and rat respond to mechanical and thermal nociceptive stimuli applied to the skin (18, 29, 32, 63, 80). Some STT cells respond

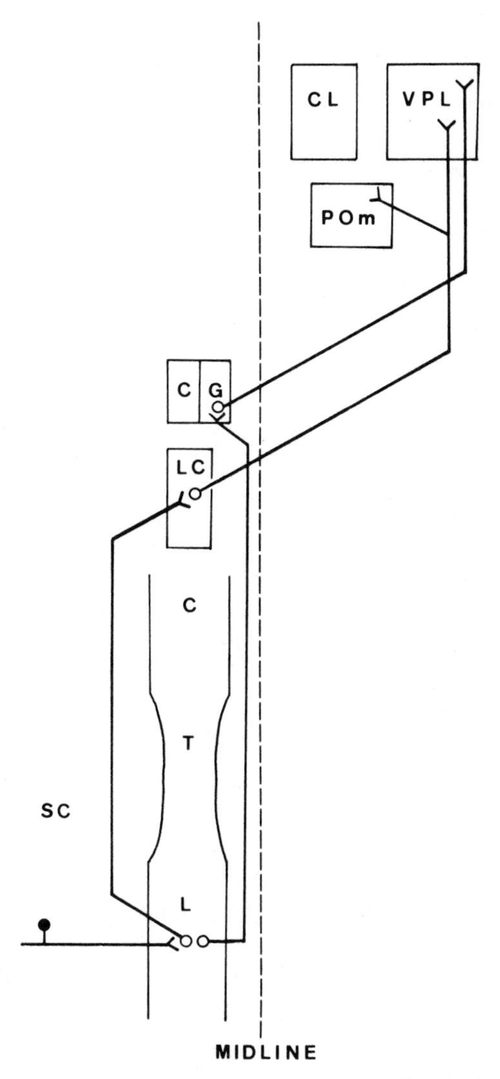

Fig. 4. Organization of candidate nociceptive pathways in dorsal white matter of spinal cord. Spinocervical (SC) and second-order dorsal column (2nd ODC) pathways from lumbosacral (L) cord are shown. Comparable pathways also originate from other levels of spinal cord, including cervical enlargement (C). The SC relays in lateral cervical (LC) nucleus. Second-order dorsal column pathway from lumbosacral enlargement relays in nucleus gracilis (G), and cervical cord pathway relays in nucleus cuneatus (C). Relay nuclei project to ventral posterior lateral (VPL) nucleus of thalamus, and cervicothalamic tract (at least) projects to medial part of posterior group (PO_m).

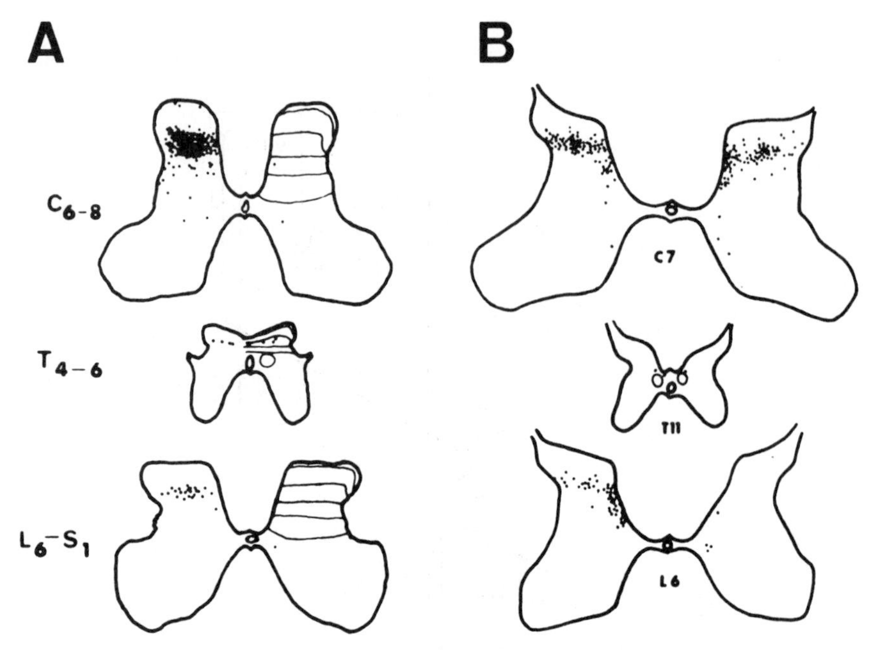

Fig. 5. Cells of origin of spinocervical tract and second-order dorsal column pathway in cat. *A*: cells labeled retrogradely by HRP injected into lateral cervical nucleus on one side. [From Craig (20).] *B*: cells labeled after injection of HRP into cuneate nuclei bilaterally and gracile nucleus on one side. [From Rustioni and Kaufman (68).]

exclusively to nociceptive stimuli (Fig. 6*A*). These are termed *high-threshold* or *nociceptive-specific cells*. Other, more abundant STT cells respond to innocuous tactile stimuli, but they respond better to nociceptive stimuli (Fig. 6*B*). These are called *wide-dynamic-range cells*, although this term is not completely satisfactory. Not germane to this discussion are STT cells that respond selectively to tactile or to proprioceptive stimuli [*low-threshold* and *deep cells* (see 29, 57, 62, 80)]. Many of these STT cells, especially those of the wide-dynamic-range category, can also be activated by noxious stimulation of muscle or viscera (Fig. 7; 26, 56).

Spinothalamic tract cells projecting to the VPL nucleus in the monkey have excitatory receptive fields on relatively restricted regions of the body surface. The receptive fields of cells in the lumbosacral enlargement are usually restricted to the ipsilateral hindlimb (32). Intense stimulation of other regions of the body surface, including the face, often results in inhibition of the discharges of the STT cell (Fig. 6; 27). The excitatory receptive fields of high-threshold STT cells are smaller than those of wide-dynamic-range STT cells (Fig. 6; 4, 62). Thus high-threshold STT cells provide a relatively unambiguous signal

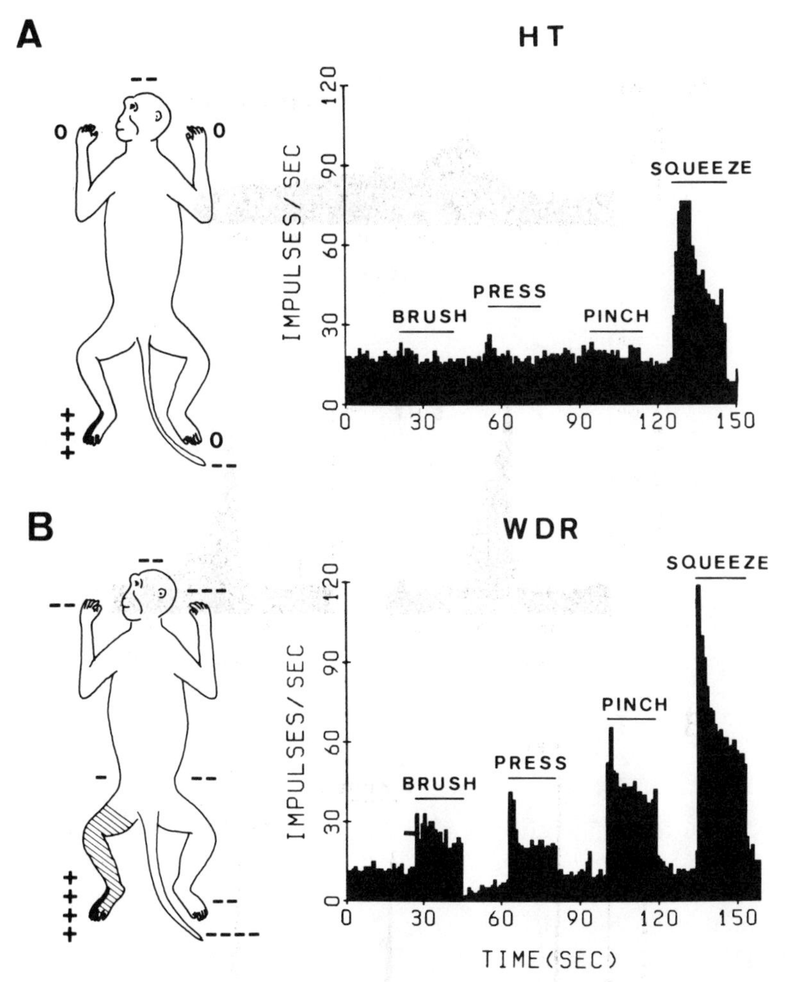

Fig. 6. Cutaneous receptive fields of spinothalamic tract cells in monkey. *A*: receptive-field distribution (figurine) and histogram (*right*) of responses of high-threshold (HT) spinothalamic tract cell. +, Excitatory receptive field (*black area* on foot); −, inhibitory receptive fields; 0, areas from which responses could not be elicited. Single-pass peri-stimulus-time histogram shows lack of response to brushing skin (brush), application of arterial clips to skin (press, pinch), and large response to squeezing skin with forceps (squeeze). *B*: receptive-field properties of wide-dynamic-range (WDR) spinothalamic tract cell. Note large receptive field with lower-threshold central zone (*black area* on foot) and surrounding higher-threshold region (*hatched area*). Inhibitory receptive fields were prominent. Histogram shows that cells responded to all gradations of stimulus intensity. [From Willis (76).]

for the location of nociceptive stimuli applied to a specific part of the body surface. Wide-dynamic-range neurons provide a more ambiguous message about the nature of the effective stimulus as well as of the

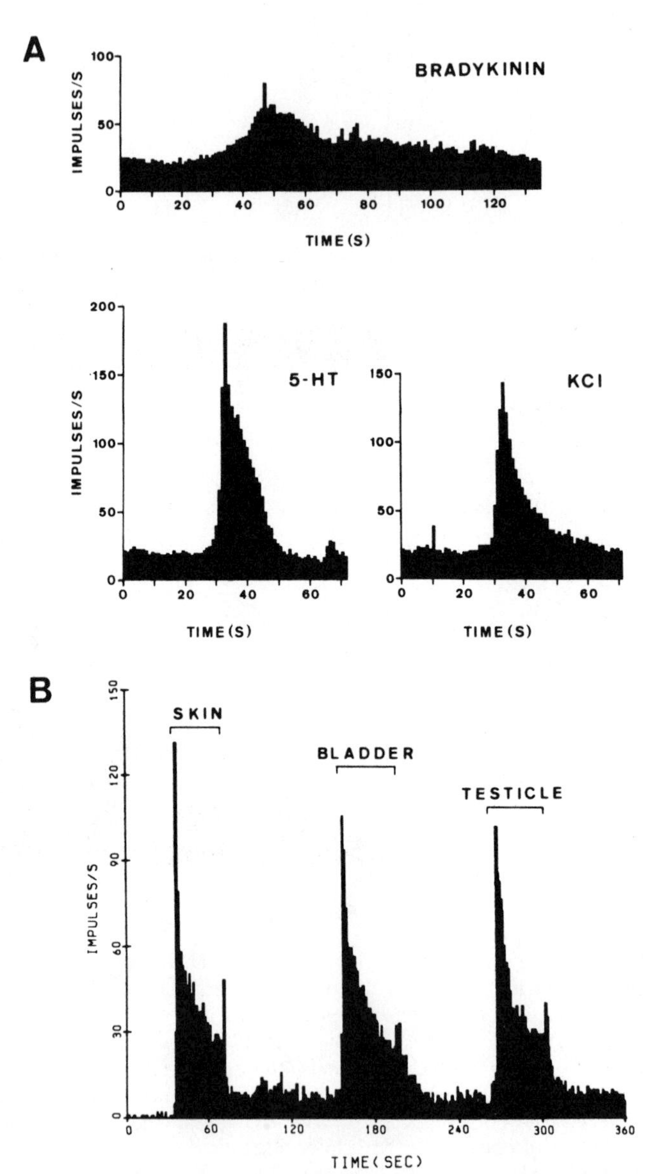

Fig. 7. Responses of spinothalamic neurons in monkey to stimulation of muscle and visceral nociceptors. *A*: histograms show averaged responses of 10 spinothalamic cells to intra-arterial injections of bradykinin, serotonin (5-HT), and KCl at doses that maximally excite fine muscle afferent fibers. [From Foreman et al. (26).] *B*: histogram shows responses of single spinothalamic tract cell to stimulation of nociceptors in skin, bladder, and testicle. [From Milne et al. (56).]

location of the stimulus. Not only is there potential for misinterpretation of the location of cutaneous stimuli, but the parallel input from nociceptors in muscle and viscera could contribute to confusion in identifying the source of input from quite separate organs. Thus wide-dynamic-range cells provide a neural mechanism that could help explain pain referral (56).

Some STT cells projecting to the VPL nucleus give off a branch to the intralaminar complex. The response properties of such cells are identical to those of STT cells projecting solely to the VPL nucleus (32).

A subpopulation of STT cells projects from the lumbosacral enlargement only to the region of the intralaminar nuclei. These cells have quite different response properties from those of STT cells projecting to the VPL nucleus or to both the VPL and the intralaminar complex (32). Most of the medially projecting cells studied so far have been classified as high-threshold cells. They respond not only to stimulation of the ipsilateral hindlimb but also to intense stimulation of much of the remaining surface of the body and face (Fig. 8). These very large

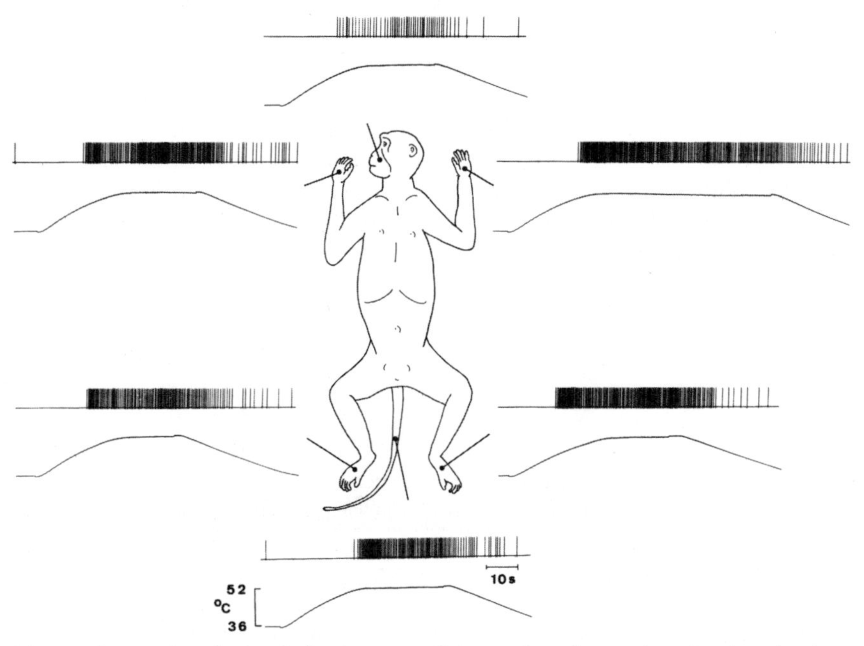

Fig. 8. Responses of spinothalamic tract cell in monkey that projected to intralaminar region of thalamus but not to ventral posterior lateral nucleus. Noxious heat stimulus applied to each point indicated on body surface. Discharges evoked by stimuli are shown on pen recordings of window discriminator pulses triggered by spikes of neuron (*upper records*). Time courses of thermal stimuli are also shown (*lower records*). Calibration shown at *bottom*. [From Giesler et al. (32).]

excitatory receptive fields depend on a neural pathway that involves a supraspinal relay, because transection of the rostral spinal cord abolishes the excitatory effects from stimulation everywhere except on the ipsilateral hindlimb (32). Cells of this type are unlikely to provide useful information about stimulus location, but they may be able to signal the presence of a noxious stimulus and to activate neurons in the medial thalamus that trigger motivational-affective reactions, including behavioral arousal.

Spinomesencephalic and spinoreticular tracts

There have only been a few studies of the response properties of spinomesencephalic and spinoreticular tract cells. A few spinothalamic tract cells projecting to the VPL nucleus in the monkey have been antidromically activated by stimulation in the midbrain in or near the lateral periaqueductal gray (62). Not surprisingly, these cells had the same response properties as other STT cells.

Menétrey et al. (55) recorded spinal neurons projecting to the midbrain reticular formation in the rat. These cells were chiefly located in the dorsal horn in areas also containing numerous STT cells, although some were in the lateral spinal nucleus. The responses of the dorsal horn cells resembled those of rat STT cells (cf. 29). However, Menétrey et al. (55) were unable to activate most of these neurons antidromically from the thalamus, and many had ipsilateral projections (unlike most STT cells). The cells therefore presumably constitute a population separate from the STT. Spinomesencephalic neurons in the lateral spinal nucleus were generally activated by stimulation of subcutaneous structures.

Spinoreticular cells projecting to the lower brain stem have been studied in the monkey (36) and the cat (24, 25, 52). Some monkey spinoreticular neurons had receptive-field properties like those of STT cells that project to the VPL nucleus; in fact some of the neurons sampled could be activated antidromically from both the medullary reticular formation and the VPL nucleus. However, other spinoreticular neurons had large, often bilateral receptive fields, more like those of STT cells projecting to the medial thalamus. Finally, a large proportion of primate spinoreticular neurons could not be activated by stimulation of the body surface (36). Most spinoreticular cells in the lumbosacral enlargement projected to the contralateral reticular formation, whereas spinoreticular cells in the cervical enlargement projected either contralaterally or ipsilaterally (cf. 48).

Spinoreticular neurons in the cat could be subdivided into cells with either restricted or extensive receptive fields (24). Many cells were best activated by noxious stimuli, whereas others had no apparent excitatory receptive field. Projections could be to either the contralateral or the ipsilateral reticular formation or both (24, 25), although in one study most projections were contralateral (52). Some spinoreticular cells in the cat could be activated antidromically from the midbrain as well as from the pontomedullary reticular formation (24). Maunz et al. (52) were less confident than Fields et al. (24) that spinoreticular cells are primarily involved in nociception.

Spinocervical tract

Although many cells of the cat spinocervical tract respond only to tactile stimuli applied to very restricted receptive fields, some have a nociceptive input (Fig. 9; 13, 14, 17, 39). Neuronal responses in the lateral cervical nucleus to pressure or even to noxious stimuli have been reported (22, 31, 35, 40, 50), but this observation should be examined further.

Second-order dorsal column pathway

Recordings have been made in the cat from neurons of both the cervical and the lumbosacral enlargements that project to the dorsal column nuclei via the dorsal funiculus (1, 2, 72). The responses of these cells are remarkably like those reported for spinothalamic neurons in the monkey and rat (Fig. 10; 18, 29, 80). Transection of the dorsal part of the cord reduces pain responses somewhat in monkeys (73) and in cats (16, 43), but it is unclear if this is due to interruption of the second-order dorsal column pathway. Apparently no studies have been done of the response properties of cells in the second-order dorsal column pathway in monkeys or rats.

Conclusions

Behavioral, anatomical, and electrophysiological evidence in monkeys, cats, and rats is consistent with a role for the following somatosensory pathways in transmitting nociceptive information from the spinal cord to the brain: spinothalamic, spinomesencephalic, spinoreticular, spinocervical, and second-order dorsal column pathways. The relative importance of these pathways in nociception appears to vary with species and circumstances. For instance, ventrally placed pathways, such as the spinothalamic and spinoreticular tracts, are probably more important for nociception in monkeys than in cats. Dorsally situated tracts may be more important in cats. Long tracts may not be as important as propriospinal pathways in the rat. When particular nociceptive pathways are interrupted, however, nociception is commonly reduced only temporarily and returns after a delay period.

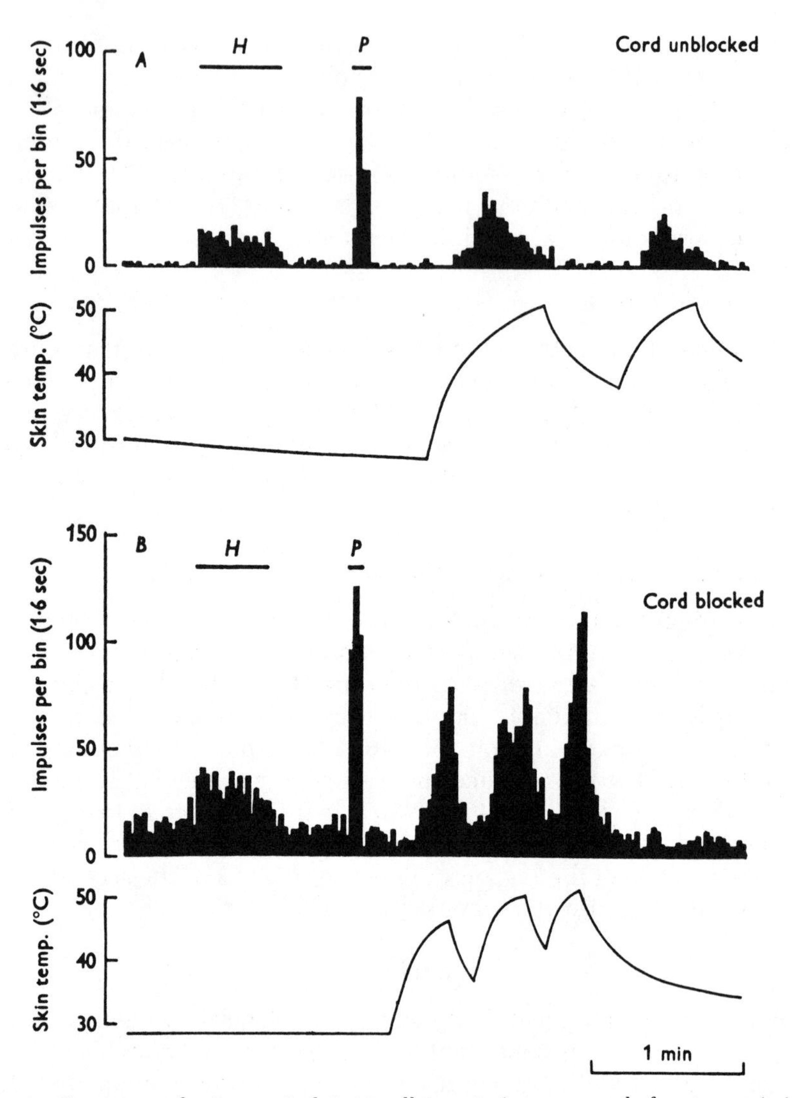

Fig. 9. Responses of spinocervical tract cell in cat. *A*: responses before transmission along spinal cord in descending inhibitory pathways was blocked with cold. Cell responded to hair movement (H), pinprick (P), and noxious heat (activity above deflections in temperature monitor). Responses were enhanced in *B* after cord was blocked. [From Cervero et al. (17).]

Presumably the return of nociceptive function is unrelated to neural regeneration but rather is due to the assumption of a significant nociceptive signaling function by the uninterrupted pathways. The dorsally located nociceptive pathways could play a role in the recovery

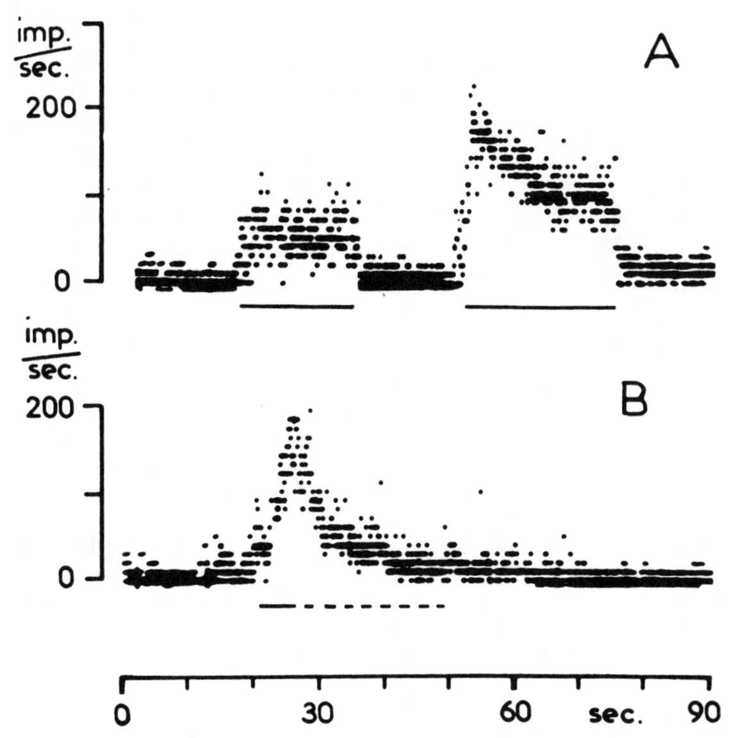

Fig. 10. Responses of cell in second-order dorsal column pathway in cat to noxious stimuli. *A*: noxious mechanical stimuli; *B*: noxious heat. [From Angaut-Petit (2).]

of nociception after cordotomy. However, the experiments of Vierck and Luck (74) provide evidence against this for the monkey.

Nociceptive signals are presumably used by various species to trigger both sensory-discriminative and motivational-affective processes (61). Pathways that appear most likely to provide useful information for the sensory-discriminative aspects of the pain response include the VPL component of the STT and the spinomesencephalic, spinocervical, and second-order dorsal column pathways. Pathways that seem better suited to a motivational-affective function are the intralaminar components of the STT and the spinoreticular tract. However, some spinothalamic cells projecting to the intralaminar region and some spinoreticular neurons have restricted receptive fields and other properties suitable for a sensory-discriminative function. Thus the intralaminar nuclei and the reticular formation cannot be ruled out as having a sensory-discriminative function on the basis of the response properties of neurons projecting to these regions.

A number of sensory phenomena related to pain in humans can be partly explained by the response properties of monkey STT cells. These phenomena include the stimulus-response curve for pain, primary and secondary hyperalgesia, referred pain, temporal summation of pain, and analgesia (cf. 18, 44, 45, 56, 77). Studies should be done to determine the possible contribution of other pathways to these pain phenomena. For example, Hong et al. (39) recently found a convergence of nociceptive input onto spinocervical tract cells in the cat from skin and from nociceptors of muscle, suggesting that such cells could contribute to the phenomenon of referred pain. As these studies are conducted, the neural bases for the species differences noted previously may become more evident.

The experiments in my laboratory were done in collaboration with the following colleagues and students: R. N. Bryan, J. M. Chung, J. D. Coulter, R. D. Foreman, K. D. Gerhart, G. J. Giesler, L. H. Haber, D. R. Kenshalo, Jr., G. A. Kevetter, R. B. Leonard, R. F. Martin, R. A. Maunz, R. J. Milne, B. D. Moore, R. F. Schmidt, H. R. Spiel, D. L. Trevino, and R. P. Yezierski. The expert technical assistance of Gail Silver, Helen Willcockson, and Grazelda Gonzales is gratefully acknowledged, as is the help of Phyllis Waldrop for typing the manuscript.

The work in my laboratory was supported by grants and fellowships from the National Institutes of Health (NS-09743, NS-11255, NS-07185, NB-43367, NS-31230, NS-05087, NS-05434, NS-05698, NS-06071, and NS-06193) and by a grant from the Moody Foundation.

REFERENCES

1. Angaut-Petit, D. The dorsal column system. I. Existence of long ascending postsynaptic fibres in the cat's fasciculus gracilis. *Exp. Brain Res.* 22: 457–470, 1975.
2. Angaut-Petit, D. The dorsal column system. II. Functional properties and bulbar relay of the postsynaptic fibres of the cat's fasciculus gracilis. *Exp. Brain Res.* 22: 471–493, 1975.
3. Antonetty, C. M., and K. E. Webster. The organization of the spinotectal projection. An experimental study in the rat. *J. Comp. Neurol.* 163: 449–466, 1975.
4. Applebaum, A. E., J. E. Beall, R. D. Foreman, and W. D. Willis. Organization and receptive fields of primate spinothalamic tract neurons. *J. Neurophysiol.* 38: 572–586, 1975.
5. Basbaum, A. I. Conduction of the effects of noxious stimulation by short-fiber multisynaptic systems of the spinal cord in the rat. *Exp. Neurol.* 40: 699–716, 1973.
6. Berkley, K. Spatial relationships between the terminations of somatic sensory and motor pathways in the rostral brainstem of cats and monkeys. I. Ascending somatic sensory inputs to lateral diencephalon. *J. Comp. Neurol.* 193: 283–317, 1980.
7. Boivie, J. The termination of the spinothalamic tract in the cat. An experimental study with silver impregnation methods. *Exp. Brain Res.* 12: 331–353, 1971.
8. Boivie, J. An anatomical reinvestigation of the termination of the spinothalamic tract in the monkey. *J. Comp. Neurol.* 186: 343–370, 1979.
9. Boivie, J. Thalamic projections from lateral cervical nucleus in monkey. A degeneration study. *Brain Res.* 198: 12–36, 1980.
10. Bowsher, D. The termination of secondary somatosensory neurons within the thalamus of *Macaca mulatta*: an experimental degeneration study. *J. Comp. Neurol.* 117: 213–227, 1961.

11. Breazile, J. E., and R. L. Kitchell. Ventrolateral spinal cord afferents to the brain stem in the domestic pig. *J. Comp. Neurol.* 133: 363–372, 1968.
12. Breazile, J. E., and R. L. Kitchell. A study of the fiber systems within the spinal cord of the domestic pig that subserve pain. *J. Comp. Neurol.* 133: 373–382, 1968.
13. Brown, A. G. Effects of descending impulses on transmission through the spinocervical tract. *J. Physiol. London* 219: 103–125, 1971.
14. Brown, A. G., and D. N. Franz. Responses of spinocervical tract neurones to natural stimulation of identified cutaneous receptors. *Exp. Brain Res.* 7: 231–249, 1969.
15. Carstens, E., and D. L. Trevino. Laminar origins of spinothalamic projections in the cat as determined by the retrograde transport of horseradish peroxidase. *J. Comp. Neurol.* 182: 151–166, 1978.
16. Casey, K. L., B. R. Hall, and T. J. Morrow. Effect of spinal cord lesions on responses of cats to thermal pulses (Abstract). *Pain* 11, Suppl.: S130, 1981.
17. Cervero, F., A. Iggo, and V. Molony. Responses of spinocervical tract neurones to noxious stimulation of the skin. *J. Physiol. London* 267: 537–558, 1977.
18. Chung, J. M., D. R. Kenshalo, Jr., K. D. Gerhart, and W. D. Willis. Excitation of primate spinothalamic neurons by cutaneous C-fiber volleys. *J. Neurophysiol.* 42: 1354–1369, 1979.
19. Corvaja, N., I. Grofova, O. Pompeiano, and F. Walberg. The lateral reticular nucleus in the cat. I. An experimental anatomical study of its spinal and supraspinal afferent connections. *Neuroscience* 2: 537–553, 1977.
20. Craig, A. D. Spinocervical tract cells in cat and dog, labeled by retrograde transport of horseradish peroxidase. *Neurosci. Lett.* 3: 173–177, 1976.
21. Craig, A. D., and H. Burton. Spinal and medullary lamina I projection to nucleus submedius in medial thalamus: a possible pain center. *J. Neurophysiol.* 45: 443–466, 1981.
22. Craig, A. D., and D. N. Tapper. Lateral cervical nucleus in the cat: functional organization and characteristics. *J. Neurophysiol.* 41: 1511–1534, 1978.
23. Ebbesson, S. O. E. Ascending axon degeneration following hemisection of the spinal cord in the Tegu lizard (*Tupinambis nigropunctatus*). *Brain Res.* 5: 178–206, 1967.
24. Fields, H. L., C. H. Clanton, and S. D. Anderson. Somatosensory properties of spinoreticular neurons in the cat. *Brain Res.* 120: 49–66, 1977.
25. Fields, H. L., G. M. Wagner, and S. D. Anderson. Some properties of spinal neurons projecting to the medial brain-stem reticular formation. *Exp. Neurol.* 47: 118–134, 1975.
26. Foreman, R. D., R. F. Schmidt, and W. D. Willis. Effects of mechanical and chemical stimulation of fine muscle afferents upon primate spinothalamic tract cells. *J. Physiol. London* 286: 215–231, 1979.
27. Gerhart, K. D., R. P. Yezierski, G. J. Giesler, and W. D. Willis. Inhibitory receptive fields of primate spinothalamic tract cells. *J. Neurophysiol.* 46: 1309–1325, 1981.
28. Giesler, G. J., D. Menétrey, and A. I. Basbaum. Differential origins of spinothalamic tract projections to medial and lateral thalamus in the rat. *J. Comp. Neurol.* 184: 107–126, 1979.
29. Giesler, G. J., D. Menétrey, G. Guilbaud, and J. M. Besson. Lumbar cord neurons at the origin of the spinothalamic tract in the rat. *Brain Res.* 118: 320–324, 1976.
30. Giesler, G. J., H. R. Spiel, and W. D. Willis. Organization of spinothalamic tract axons within the rat spinal cord. *J. Comp. Neurol.* 195: 243–252, 1981.
31. Giesler, G. J., G. Urca, J. T. Cannon, and J. C. Liebeskind. Response properties of neurons of the lateral cervical nucleus in the rat. *J. Comp. Neurol.* 186: 65–78, 1979.
32. Giesler, G. J., R. P. Yezierski, K. D. Gerhart, and W. D. Willis. Spinothalamic tract neurons that project to medial and/or lateral thalamic nuclei: evidence for a physiologically novel population of spinal cord neurons. *J. Neurophysiol.* 46: 1285–1308, 1981.

33. Gwyn, D. G., and H. A. Waldron. A nucleus in the dorsolateral funiculus of the spinal cord of the rat. *Brain Res.* 10: 342–351, 1968.
34. Ha, H. Cervicothalamic tract in the Rhesus monkey. *Exp. Neurol.* 33: 205–212, 1971.
35. Ha, H., S. T. Kitai, and F. Morin. The lateral cervical nucleus of the racoon. *Exp. Neurol.* 11: 441–450, 1965.
36. Haber, L. H., B. D. Moore, and W. D. Willis. Electrophysiological response properties of spinoreticular neurons in the monkey. *J. Comp. Neurol.* 207: 75–84, 1982.
37. Hardy, J. D., H. G. Wolff, and H. Goodell. *Pain Sensations and Reactions.* Baltimore, MD: Williams & Wilkins, 1952. [Reprinted by Hafner, New York, 1967.]
38. Hazlett, J. C., R. Dom, and G. F. Martin. Spino-bulbar, spino-thalamic and medial lemniscal connections in the American opossum, *Didelphis marsupialis virginiana. J. Comp. Neurol.* 146: 95–118, 1972.
39. Hong, S. K., K. D. Kniffki, S. Mense, R. F. Schmidt, and M. Wendisch. Descending influences on the responses of spinocervical tract neurones to chemical stimulation of fine muscle afferents. *J. Physiol. London* 290: 129–140, 1979.
40. Horrobin, D. F. The lateral cervical nucleus of the cat; an electrophysiological study. *Q. J. Exp. Physiol.* 51 351–371, 1966.
41. Jones, E. G., and H. Burton. Cytoarchitecture and somatic sensory connectivity of thalamic nuclei other than the ventrobasal complex in the cat. *J. Comp. Neurol.* 154: 395–432, 1974.
42. Karten, H. J. Ascending pathways from the spinal cord in the pigeon (*Columba livia*). *Proc. Int. Congr. Zool., 16th, Washington, DC, 1963,* p. 23.
43. Kennard, M. A. The course of ascending fibers in the spinal cord of the cat essential to the recognition of painful stimuli. *J. Comp. Neurol.* 100: 511–524, 1954.
44. Kenshalo, D. R., Jr., R. B. Leonard, J. M. Chung, and W. D. Willis. Responses of primate spinothalamic neurons to graded and to repeated noxious heat stimuli. *J. Neurophysiol.* 42: 1370–1389, 1979.
45. Kenshalo, D. R., Jr., R. B. Leonard, J. M. Chung, and W. D. Willis. Facilitation of the responses of primate spinothalamic cells to cold and to tactile stimuli by noxious heating of the skin. *Pain* 12: 141–152, 1982.
46. Kerr, F. W. L. The ventral spinothalamic tract and other ascending systems of the ventral funiculus of the spinal cord. *J. Comp. Neurol.* 159: 335–356, 1975.
47. Kerr, F. W. L., and H. H. Lippman. The primate spinothalamic tract as demonstrated by anterolateral cordotomy and commissural myelotomy. *Adv. Neurol.* 4: 147–156, 1974.
48. Kevetter, G. A., L. H. Haber, R. P. Yezierski, J. M. Chung, R. F. Martin, and W. D. Willis. Cells of origin of the spinoreticular tract in the monkey. *J. Comp. Neurol.* 207: 61–74, 1982.
49. Kevetter, G. A., and W. D. Willis. Spinothalamic cells in the rat lumbar cord with collaterals to the medullary reticular formation. *Brain Res.* 238: 181–185, 1982.
50. Kitai, S. T., H. Ha, and F. Morin. Lateral cervical nucleus of the dog: anatomical and microelectrode studies. *Am. J. Physiol.* 209: 307–312, 1965.
51. Lund, R. D., and K. E. Webster. Thalamic afferents from the spinal cord and trigeminal nuclei. An experimental anatomical study in the rat. *J. Comp. Neurol.* 130: 313–328, 1967.
52. Maunz, R. A., N. G. Pitts, and B. W. Peterson. Cat spinoreticular neurons: locations, responses and changes in responses during repetitive stimulation. *Brain Res.* 148: 365–379, 1978.
53. Mehler, W. R. Some neurological species differences—a posteriori. *Ann. NY Acad. Sci.* 167: 424–468, 1969.
54. Mehler, W. R., M. E. Feferman, and W. J. H. Nauta. Ascending axon degeneration

following anterolateral cordotomy. An experimental study in the monkey. *Brain* 83: 718–751, 1960.
55. Menétrey, D., A. Chaouch, and J. M. Besson. Location and properties of dorsal horn neurons at origin of spinoreticular tract in lumbar enlargement of the rat. *J. Neurophysiol.* 44: 862–877, 1980.
56. Milne, R. J., R. D. Foreman, G. J. Giesler, and W. D. Willis. Convergence of cutaneous and pelvic visceral nociceptive inputs onto primate spinothalamic neurons. *Pain* 11: 163–183, 1981.
57. Milne, R. J., R. D. Foreman, and W. D. Willis. Responses of primate spinothalamic neurons located in the sacral intermediomedial gray (Stilling's nucleus) to proprioceptive input from the tail. *Brain Res.* 234: 227–236, 1982.
58. Mizuno, N., K. Nakano, M. Imaizumi, and M. Okamoto. The lateral cervical nucleus of the Japanese monkey (*Macaca fuscata*). *J. Comp. Neurol.* 129: 375–384, 1967.
59. Morin, F. A new spinal pathway for cutaneous impulses. *Am. J. Physiol.* 183: 245–252, 1955.
60. Morin, F., H. G. Schwartz, and J. L. O'Leary. Experimental study of the spinothalamic and related tracts. *Acta Psychiatr. Neurol.* 26: 371–396, 1951.
61. Price, D. D., and R. Dubner. Neurons that subserve the sensory discriminative aspects of pain. *Pain* 3: 307–338, 1977.
62. Price, D. D., R. L. Hayes, M. Ruda, and R. Dubner. Spatial and temporal transformation of input to spinothalamic tract neurons and their relations to somatic sensations. *J. Neurophysiol.* 41: 933–947, 1978.
63. Rao, G. S., J. E. Breazile, and R. L. Kitchell. Distribution and termination of spinoreticular afferents in the brain stem of sheep. *J. Comp. Neurol.* 137: 185–196, 1969.
64. Rexed, B. The cytoarchitectonic organization of the spinal cord in the cat. *J. Comp. Neurol.* 96: 415–495, 1952.
65. Rexed, B., and A. Brodal. The nucleus cervicalis lateralis. A spino-cerebellar relay nucleus. *J. Neurophysiol.* 14: 399–407, 1951.
66. Rustioni, A. Non-primary afferents to the nucleus gracilis from the lumbar cord of the cat. *Brain Res.* 51: 81–95, 1973.
67. Rustioni, A. Non-primary afferents to the cuneate nucleus in the brachial dorsal funiculus of the cat. *Brain Res.* 75: 247–259, 1974.
68. Rustioni, A., and A. B. Kaufman. Identification of cells of origin of non-primary afferents to the dorsal column nuclei of the cat. *Exp. Brain Res.* 27: 1–14, 1977.
69. Trevino, D. L. The origin and projections of a spinal nociceptive and thermoreceptive pathway. In: *Sensory Functions of the Skin in Primates, With Special Reference to Man*, edited by Y. Zotterman. New York: Pergamon, 1976, p. 367–376.
70. Trevino, D. L., and E. Carstens. Confirmation of the location of spinothalamic neurons in the cat and monkey by the retrograde transport of horseradish peroxidase. *Brain Res.* 98: 177–182, 1975.
71. Trevino, D. L., R. A. Maunz, R. N. Bryan, and W. D. Willis. Location of cells of origin of the spinothalamic tract in the lumbar enlargement of cat. *Exp. Neurol.* 34: 64–77, 1972.
72. Uddenberg, N. Functional organization of long, second-order afferents in the dorsal funiculus. *Exp. Brain Res.* 4: 377–382, 1968.
73. Vierck, C. J., D. M. Hamilton, and J. I. Thornby. Pain reactivity of monkeys after lesions to the dorsal and lateral columns of the spinal cord. *Exp. Brain Res.* 13: 140–158, 1971.
74. Vierck, C. J., and M. M. Luck. Loss and recovery of reactivity to noxious stimuli in monkeys with primary spinothalamic cordotomies, followed by secondary and tertiary lesions of other cord sectors. *Brain* 102: 233–248, 1979.

75. White, J. C., and W. H. Sweet. *Pain and the Neurosurgeon.* Springfield, IL: Thomas, 1969.
76. Willis, W. D. Ascending pathways from the dorsal horn. In: *Spinal Cord Sensation*, edited by A. G. Brown and M. Réthelyi. Edinburgh: Scottish Academic, 1981, p. 169–178.
77. Willis, W. D. The spinothalamic tract. In: *The Clinical Neurosciences*, edited by R. H. Rosenberg. New York: Churchill Livingstone, sect. V, in press.
78. Willis, W. D., and R. E. Coggeshall. *Sensory Mechanisms of the Spinal Cord.* New York: Plenum, 1978.
79. Willis, W. D., D. R. Kenshalo, Jr., and R. B. Leonard. The cells of origin of the primate spinothalamic tract. *J. Comp. Neurol.* 188: 543–574, 1979.
80. Willis, W. D., D. L. Trevino, J. D. Coulter, and R. A. Maunz. Responses of primate spinothalamic tract neurons to natural stimulation of hindlimb. *J. Neurophysiol.* 37: 358–372, 1974.
81. Yoss, R. E. Studies of the spinal cord. Pt. 3. Pathways for deep pain within the spinal cord and brain. *Neurology* 3: 163–175, 1953.

FOUR

Supraspinal Pain Mechanisms in the Cat

Kenneth L. Casey
Thomas J. Morrow
Neurology Research Laboratories, Veterans Administration Medical Center, Ann Arbor, Michigan

The participants at the Neurosciences Research Program (NRP) work session on pain (64) suggested some lines of evidence that would be useful for identifying neurons that mediate pain. It is necessary to specify these lines of evidence because noxious stimuli elicit somatic and autonomic reflexes, somatic or visceral sensations, and strong motivational and affective experiences. Neurons responding differentially or exclusively to noxious stimuli might mediate only one of these effects. For example, nociceptive neurons could encode information about the timing, location, and intensity of the noxious stimulus without mediating the aversive affective experience that is an essential component of pain (93). The identification of pain-mediating neurons is further complicated because some central nervous system neurons responding to noxious stimuli may suppress rather than initiate pain behavior and the nociceptive responses of other neurons (4, 66, 70).

In reviewing what is known about supraspinal pain pathways and mechanisms in animals, we have found it helpful to refer to the lines of evidence suggested at the NRP work session. We list them here [for more detail see original publication (64); Price and Dubner (88) have more completely elaborated similar lines of evidence pertaining to the sensory-discriminative aspects of pain].

1) *Neuronal responses.* Candidate neurons should respond exclusively or differentially (i.e., with a higher frequency) to noxious somatic or visceral stimuli.

2) Attenuation. Reducing the responses of candidate neurons to noxious stimuli should reduce pain.

3) Stimulation. Selective stimulation of candidate neurons should elicit pain-related experience or behavior.

4) Hodology. Candidate neurons should have anatomical connections with neurons that, by other lines of evidence, probably contribute to pain.

We are at a very early stage in establishing these lines of evidence for pain mechanisms in animals. Price and Dubner (88) have reviewed the case for sensory-discriminative peripheral and central neurons subserving pain in animals and humans. Other reviews covering supraspinal pain pathways and mechanisms have appeared recently (25, 64, 100). In this chapter, we concentrate on the evidence as it pertains to supraspinal mechanisms in the cat, presenting some results from our present studies and referring to related studies in the human and monkey.

Nocifensive Responses in Cat

A significant impediment to establishing these lines of evidence in animals is the few behavioral tests for pain, especially in the cat. Behavioral tests are essential for establishing evidence for attenuation and stimulation of neurons mediating pain and are necessary to determine the types of stimuli that are noxious.

For our studies in the cat, we use thermal pulse stimuli to deliver a natural, well-controlled, temporally discrete stimulus that excites a limited number of physiologically specific afferent fibers (16). We use a water-cooled, spring-loaded contact thermode (75). It is important to have a rapid pulse onset (up to $40°C \cdot s^{-1}$) and rapid, active cooling when the cat's responses terminate the pulse. The thermodes, set at an adapting temperature of 38°C, are placed against the shaved outer thighs of partially restrained cats that have been trained to place their muzzles into a food cup that is periodically illuminated to signal the availability of pureed cat food. The cat's muzzle interrupts a photocell contact above the food cup; this triggers food delivery and, on a semirandom schedule assuring a >3-min interval between stimuli at any site, initiates thermal pulse stimuli of various intensities. The occurrence and latency of three behavioral responses are detected by appropriate electronic devices during the 7-s stimulus period: interruption of eating (muzzle withdrawal), hindlimb movement, and vocalization.

Casey et al. (27) have shown that the probability of each of these behavioral responses increases with thermal stimulus intensity. Figure

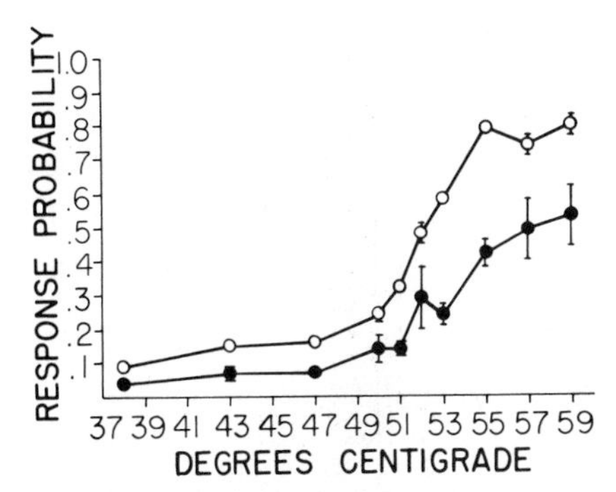

Fig. 1. Stimulus-response probability profiles of 29 cats. ○, Probability (± SEM) that any 1 of 3 nocifensive behaviors (interruption of eating, hindlimb movement, or vocalization) was elicited by thermal pulse to shaved outer thigh. ●, Average probablity profile of all 3 behaviors. Average number of trials ranged from 176 to 16, given over multiple sessions to each cat at each temperature.

1 shows, for 29 cats, the stimulus-response profiles for emitting any one or more of the behaviors (○) and the average profile of all behaviors (●). The average response threshold (above the 95% confidence limits of probability at 38°C) is 50°C, which is close to the nocifensive threshold for cat as determined by other means (62, 90). Figure 2 shows a histogram of the average response latencies of all 29 cats based on an analysis of the largest number of responses emitted in any behavioral category for each cat. The interpretation that these are natural, unlearned escape responses to noxious stimuli is supported by several observations: *1*) the responses require stimuli within the threshold range of thermal nociceptive afferents (5, 6, 11, 38, 96); *2*) there is no evidence for learning, based on an analysis of response choices and trends; *3*) responses followed the onset of the thermal-pulse plateau, and their latency was not influenced by reducing the slope of pulse onset to $17°C \cdot s^{-1}$; and *4*) these otherwise unrewarded responses showed stable probabilities over hundreds of trials and an increased probability as stimulus intensity increased. Avoidance responses to innocuous stimuli would be expected to reach a high probability at lower stimulus intensities and show little or no increase at higher stimulus strengths.

We have used this behavioral paradigm in the cat to determine if lesions at various central nervous system sites reduce nocifensive responses without affecting the motor ability as measured by response latency. In reviewing evidence for a limited number of supraspinal structures, we present some examples from our studies, showing that some lesions attenuate whereas others enhance these nocifensive behaviors.

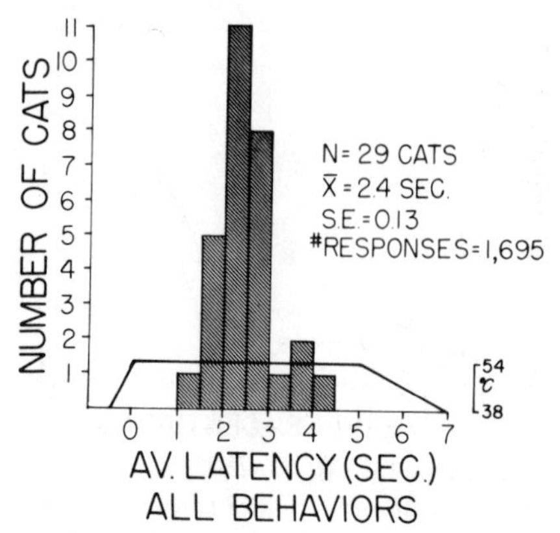

Fig. 2. Average response latencies of 29 cats, based on analysis of largest number of responses in any behavioral category for each cat. Stimulus intensities eliciting behaviors ranged from 53 to 55°C.

Brain Stem

Hodology

The medial medullary reticular formation in the region of n. reticularis gigantocellularis (RGC) and n. reticularis magnocellularis (RMC) receives input from the ventral anterior and lateral spinal funiculi in a number of species, including cat (71). Medullary reticulospinal neurons descend to the spinal cord (78). Most descending reticulospinal projections from RGC are to the ventral spinal gray, but the RMC has connections with neurons of the dorsal horn (3). Bulboreticular neurons also project rostrally to the medial and intralaminar thalamus and subthalamus (76). The medial thalamus and subthalamus of cat have been implicated in pain mechanisms based on neurophysiological (32) and behavioral (53, 72) evidence. Anatomically, then, the medial medullary reticular formation could mediate some component of pain.

In contrast, neurons of the midline medullary tegmentum, especially those of the n. raphe magnus (NRM), have been implicated in mechanisms of analgesia because of anatomical (3, 4), electrophysiological (69), and behavioral (82) evidence that their activity suppresses the responses of dorsal horn neurons to noxious stimuli (for review see 4, 37, 70).

Neuronal responses

Several studies have reported that a significant proportion of RGC, RMC, and nearby medial reticular cells respond differentially or exclusively to noxious somatic stimuli (22, 23, 41, 43, 67, 84). The large and frequently bilateral receptive fields of these neurons, however, suggest that these cells do not mediate the discriminative aspects of nociception. The observations of Siegel (94) in the awake, unrestrained cat further suggest that medial bulboreticular cells that respond to noxious stimuli may mediate motor responses.

Stimulation

Electrical stimulation within RGC and dorsal RMC has been shown to be aversive in cat (24) and rat (19, 60). This contrasts with the analgesic effects of stimulation within the NRM (82). Motor and autonomic effects of RGC and RMC stimulation have also been reported (24, 50).

Attenuation

Microinjection of opiate narcotic analgesics into medial reticular nuclei of rat has been reported to produce analgesia, but whether this procedure attenuates or enhances the activity of these cells is not clear (37). We found in the rat that the spontaneous and somatically evoked activity of RGC and RMC neurons was suppressed during analgesia (tail-flick response) produced by electrical stimulation within the periaqueductal gray (74). In the cat, electrolytic coagulation lesions within the RGC and RMC have been shown to reduce escape responses of cats to peripheral electric shock (46).

We placed thermocoagulation lesions in the medial medulla of cats to test their effect on the stimulus-response profiles to noxious thermal stimuli. The results, represented in Figure 3, are consistent with the previously cited evidence that pain responses are mediated by neurons or fibers of passage within the medial medullary reticular substance, whereas analgesia is mediated by ventral midline raphe neurons such as the NRM. Figure 3*A* shows that a midline lesion spared NRM but destroyed paramedial reticular neurons and fibers within the dorsal midline. Medial reticular structures, including the RGC, and the dorsal part of the RMC were extensively damaged by the adjacent unilateral lesion. Postoperatively this cat no longer showed an interrupt response even to stimuli more intense than those that were previously effective. Motor function was unimpaired, as determined by clinical inspection and lack of change in hindlimb movement latency. In contrast, Figure 3*B* shows markedly increased movement and interrupt responses (not shown) after more ventrally placed lesions, which destroyed nearly all neurons of the NRM and neurons within the RMC but spared substantial portions of the more dorsolateral RGC.

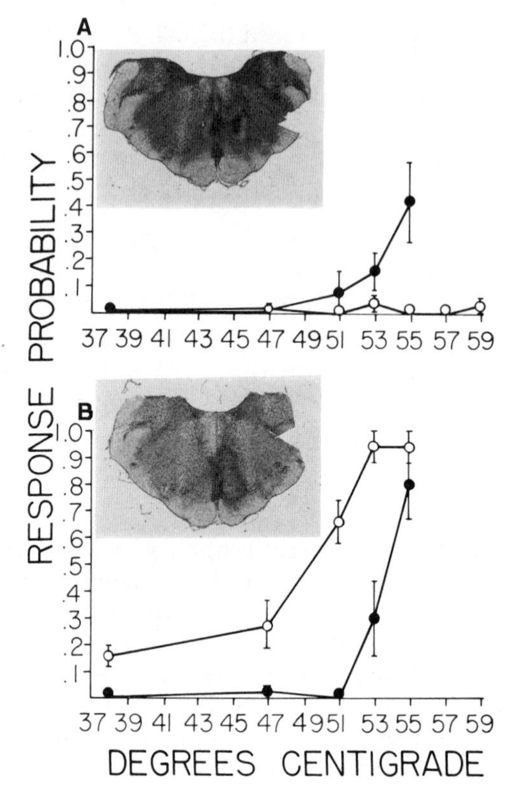

Fig. 3. Contrasting effects of bulboreticular lesions on nocifensive responses in cat. ●, Preoperative stimulus-response curves; ○, postoperative curves. *A*: lesion resulted in loss of interruption of eating elicited by stimulus. *B*: lesion produced increased interruption (not shown) and hindlimb movement responses. Response latencies of both cats were unchanged.

Summary

All lines of evidence indicate that the RGC and possibly the dorsal RMC regions mediate some dimension of pain. Whether these cells act via ascending reticulodiencephalic pathways, reticulospinal connections, or both remains to be determined. The response properties of these cells, however, make them unlikely candidates for the mediation of sensory-discriminative features of pain. Furthermore the extensive anatomical connections of these neurons, and the fact that motor as well as autonomic and aversive behavioral reactions are elicited by electrical stimulation within their territory, support Casey's (26) view that the medial medullary reticular formation mediates a wide variety of reactions that typically accompany the activation of nociceptors. These functions contrast with the nocifensive suppression mediated by the midline neurons of the NRM.

Thalamus

Hodology

The pathways that transmit somatosensory information most directly from spinal to supraspinal sites include the dorsal columns (DC), the spinocervicothalamic pathway (SCT), and the spinothalamic tract (STT). There is behavioral and neurophysiological evidence that the SCT and STT include nociceptive neurons and are necessary for normal nocifensive responses in the cat (15, 61). Nociceptive fibers have been recorded from the dorsal columns (2), but the available behavioral evidence indicates that they do not mediate nocifensive responses in the cat (35, 61).

The SCT projects to the ventral posterolateral (VPL), medial posterior (POM), and magnocellular medial geniculate (MGM) thalamic nuclei (9, 12). The STT of cat also has terminal endings in POM and MGM and sends additional terminals into the zona incerta (ZI) and the ventrolateral (VL) thalamic nucleus (9, 13). Medial and intralaminar thalamic nuclei also receive direct input from the STT [centralis lateralis (CL), paracentralis (PC), parafascicularis (PF), subparafascicularis (SPF), and centralis medialis (CEM)] but not from the SCT (9, 12, 13). Of particular interest is the submedius nucleus (SUM), which Craig and Burton (29) have shown in cat to receive direct input from neurons of the marginal zone of dorsal horn, a source of specific nociceptive input.

There is evidence that SUM neurons project to the orbitofrontal cortex of cat (30, 77). As a group, the intralaminar thalamic nuclei project diffusely to extensive areas of frontal, parietal, and medial cortex and also to the ipsilateral striatum (52). Jones and Leavitt (52) demonstrated CL projections to the sensorimotor cortex in cat and monkey. The PF and SPF nuclei also project to the amygdala in cat (83). In the lateral thalamus the cortical projections of POM and MGM have been shown by Heath (48), Heath and Jones (49), Jones and Leavitt (51), and Graybiel (42) to include the insular cortex and the regions adjacent to and overlapping the second somatosensory cortex of cat. The VL nucleus of cat projects to Brodmann's area 3a of the cortex, at the border between motor and primary somatosensory cortex (95). That region of the medial ZI thought to subserve nociceptive functions has subcortical projections that include certain midline and intralaminar thalamic nuclei, the H field of Forel, and the posterior and dorsal hypothalamus (58).

Neuronal responses

There is evidence that some nociceptive neurons are located in the ventrobasal or ventral posterolateral thalamic nucleus of the cat (44,

87). A modest proportion of VPL neurons recorded from the anesthetized monkey have small to medium-sized contralateral receptive fields and receive nociceptive input (63). Some cat VPL neurons were shown to have wide receptive fields (8, 47) and to respond to stimulation of C-fiber afferents (36), but the great majority of VPL cells in the cat respond exclusively to innocuous tactile stimuli. Recordings from the dorsal and ventral periphery of VPL in the anesthetized cat have recently revealed neurons responding exclusively or differentially to noxious stimulation of skin (50a) or muscle and tendon (64a). We have found posterior lateral VPL neurons that are differentially responsive to noxious stimulation of the skin of the awake, behaving squirrel monkey (27a). That portion of the VL nucleus receiving STT input in the cat has not been intensively studied, but Nyquist (79) found that only 6% of her sample of 244 neurons responded to somatic stimuli, and none were nociceptive.

In contrast, there are several reports of posterior thalamic (PO) neurons in the cat responding differentially or exclusively to noxious stimuli with the proportion of such neurons ranging from 58% (87) to 30% (14). Why such high proportions of nociceptive PO neurons have not been reported by others is not clear (31, 81).

There are also reports of many nociceptive cells in the PF, SPF, CL (32), ZI, and centromedian (CM) (85) nuclei of cat (see also refs. 1, 80). The cat SUM nucleus is also reported to contain nociceptive neurons (29). In the rat, cells in certain intralaminar nuclei have responded primarily to noxious stimuli (86) and have discharged maximally to aversive reticular formation stimuli (59). Neurons that are differentially but not exclusively responsive to noxious stimuli have been recorded from the PO and intralaminar nuclei of the awake squirrel monkey (21).

Stimulation

Aversive behavior can be elicited by stimulation in or near the PO nucleus in cat (7). In humans there is evidence that electrical stimulation within the posterior ventrolateral thalamus elicits painlike behavior and reports of pain (45).

Stimulation within the medial and intralaminar nuclei may elicit fearlike responses, but both arousal and sleep responses have also been observed, depending on the stimulus parameters employed (65). Kaelber and Mitchell (55) have shown, however, that stimulation within CM, PF, SPF, or the rostral central tegmental fasciculus can be used to train cats to give escape responses. In subsequent studies (53, 54, 56) it was found that electrical stimuli within the medial ZI and the H–H_2 fields of Forel are especially effective in eliciting nocifensive responses and escape behavior (barrier crossing). Stimulation in subthalamic and medial thalamic structures that project to the medial ZI and H–H_2 fields was also shown to be aversive.

Attenuation

There are no studies on the effect of PO lesions on pain in the cat. The PO lesions, unlike those within the more rostral ventrobasal complex, do not interfere with somesthetic direction discrimination in the cat (40), but there is no evidence regarding nocifensive responses. The studies of Mitchell and Kaelber (72, 73), however, suggest that the integrity of the PO is not, by itself, sufficient to maintain normal escape responses to grid shock or tooth-pulp stimulation in the cat. They found that lesions destroying most of the CM-PF-SPF complex effectively eliminated responses to tooth-pulp stimulation and interfered with or eliminated escape from grid shock. Cats that failed to escape grid shock, however, continued to display nocifensive behaviors.

In a subsequent study, Kaelber et al. (57) demonstrated that smaller CM-PF lesions could eliminate or attenuate learned escape responses to tooth-pulp shock, while sparing the ability of cats to learn avoidance of grid shock. Overall the results are consistent with the hypothesis that the spared diencephalic structures, possibly including the PO, can mediate pain from the body. Results supporting this position are shown in Figure 4*A*. The interrupt responses of this cat to noxious thermal stimulation of either hindlimb were nearly eliminated by a unilateral lesion that destroyed much of the PO region and extended into the MGM and adjacent nuclei. Response latency was unaffected by the lesion. The bilateral effects of this lesion may be due to the bilateral representation of receptive fields in this thalamic area (8, 14, 44, 87).

Medial thalamic lesions, in contrast, somewhat unexpectedly increased nocifensive responsiveness, as shown in Figure 4*B*. The significant leftward shift of this cat's stimulus-response curve followed a lesion that destroyed most of the PF and SPF nuclei and extended laterally to involve a substantial portion of CM bilaterally. Midline thalamic nuclei, such as CEM, were also destroyed. Why this result is in such marked contrast to those of Mitchell and Kaelber (72, 73) and Kaelber et al. (57) is not clear, but the involvement of more midline structures in our lesion and differences in the type of noxious stimulus and behavioral paradigm that were used may be significant. In any case, the example in Figure 4 (as in Fig. 3) emphasizes the importance of lesion analysis and shows that the complete or partial destruction of structures known to include nociresponsive neurons may enhance, rather than attenuate, nocifensive behavior. This point is made again by the example in Figure 5, where the SUM nuclei, known to receive

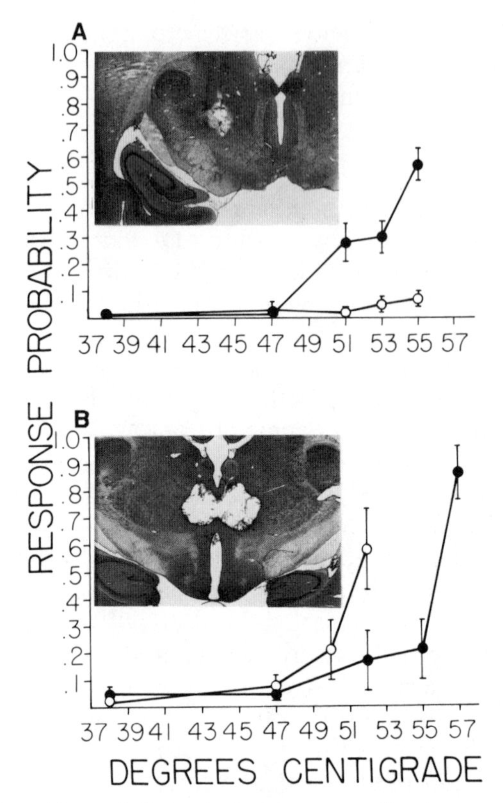

Fig. 4. Contrasting effects of thalamic lesions on nocifensive responses in cat. ●, Preoperative data; ○, postoperative data. *A*: lesion nearly eliminated interrupt responses elicited by stimulation of either hindlimb. *B*: lesion increased hindlimb-movement responses; interrupt and vocalization responses of this cat were too infrequent to show significant changes. Response latencies of both cats were unchanged.

direct nociceptive input from the cat dorsal horn (29), have been damaged extensively by bilateral lesions that resulted in a modest but significant enhancement of hindlimb-movement responses to noxious thermal stimuli. It is possible that these medial and midline thalamic lesions (Figs. 4 and 5), like the midline brain stem lesion (Fig. 3), disrupt an endogenous pain-suppression feedback mechanism in the cat, as suggested by the results of brain-stimulation studies in humans (91).

Summary

The lines of evidence for pain mediation are not clearly established for the various thalamic nuclei. The few attempts to attenuate the

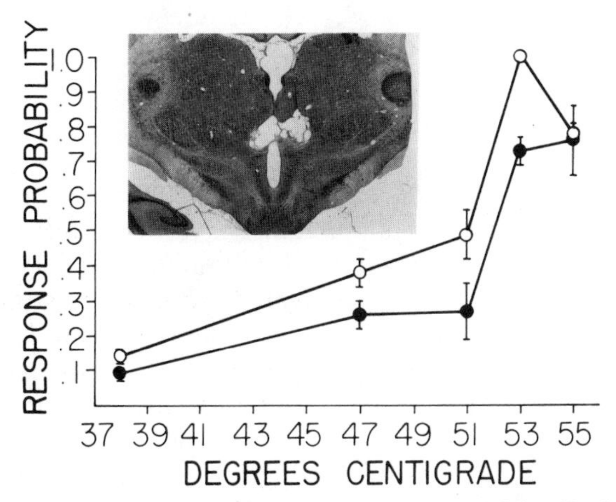

Fig. 5. Medial thalamic lesion that significantly increased hindlimb-movement responses. Other nocifensive behaviors were not significantly increased, and response latencies were unchanged. Lesion has extensively damaged CEM and adjacent SUM nuclei at levels rostral and caudal to sample shown, although portions of both nuclei have been spared.

function of anatomically distinct thalamic nuclei have raised serious questions about the functional significance of neuronal activity in the target areas. These problems can only be resolved by further experiments in this largely unexplored area of pain research.

Currently the weight of available evidence seems, in our opinion, to support the view that neurons in the posterior lateral thalamus are necessary for the normal recognition and localization of noxious somatic stimuli. Neurons within the medial and intralaminar thalamus, however, appear to mediate a variety of equally essential functions in pain mechanisms. These functions may include the initiation of arousal, aversive affective mechanisms and behavior, autonomic and somatic motor responses, and the activation of endogenous control mechanisms that attenuate responses to noxious stimuli.

Cerebral Cortex

Hodology

The thalamocortical connections with sensorimotor cortex of cat have been discussed. Presumably this region of the neocortex mediates functions relevant to the sensory-discriminative aspect of pain. In regard to motivational and affective dimensions of pain, Kaelber and Smith (58) found that fibers from the medial ZI, where electrical stimulation elicits brisk escape responses, project to midline, intralaminar, and anterior thalamic nuclei that have recently been shown to project to anterior limbic and cingular cortex (77, 92).

Neuronal responses

In the anesthetized cat (20) and monkey (98), recordings from the posterior region of the second somatosensory area (SII) and adjacent cortex, which receives PO thalamic input, have revealed a population of neurons that are primarily activated by noxious stimuli applied over wide areas of the body surface. A systematic search for nociceptive neurons in limbic cortical areas of the cat has not been reported. Evoked-potential recordings from humans suggest that the midline (vertex) potentials elicited by specific noxious stimuli reflect the activation of cortical structures mediating the arousal, alerting, and possibly affective dimensions of pain (17, 18, 28). Whether these evoked responses in turn reflect the activation of midline and intralaminar thalamic nuclei is not known.

Stimulation

Durinyan et al. (33) have reported strong alerting responses to electrical stimulation of SII in cat, but it is not known if this stimulation is aversive. Verbal reports of pain have been elicited in humans during electrical stimulation of SII and adjacent parietal cortex (99).

Attenuation

The cortical ablation experiments of Berkeley and Parmer (10) provide evidence that the SII cortex and adjacent sulci are necessary for the normal execution of escape responses (barrier crossing) to electric grid shock in the cat. However, cats that were overtrained for 6–7 mo before surgery were not affected by the cortical lesions. Glassman (39) found deficits in cutaneous tactile detection after SII ablations in cat, but nociception was not tested. Perhaps the most convincing evidence for a long-lasting and localized deficit in pain sensation attributable to a cortical lesion comes from the clinical observations of humans with cortical damage from bullet wounds (68). Unfortunately, similar subsequent reports have not appeared, and an anatomical analysis of the critical lesion(s) necessary to produce this hypalgesia is not available.

In humans, cortical lesions outside the somatosensory receiving areas have been used to relieve intractable pain without affecting somesthetic discriminative capacity. Presumably such lesions as the frontal leukotomy and cingulumotomy (34, 97) alter the affective and emotional aspects of pain. Randa (89) has reported that cats with extensive damage to the cingulum fasciculus show increased threshold and latency to escape from tooth-pulp stimulation for at least 1 mo after surgery. Damage to the cingulate gyrus with partial sparing of the cingulum fasciculus, however, resulted in a markedly reduced escape threshold.

Summary

The lines of evidence that some neurons in cat somatosensory cortex mediate pain are partially but not completely supported. There are experiments and human clinical observations that favor this hypothesis, although data are very limited and, in some cases, consist of one set of experiments. The interpretation of the results of the behavioral experiments with peripheral electric shock is somewhat complicated because such stimuli synchronously excite a wide spectrum of functionally heterogeneous cutaneous afferents. A volley of impulses in nonnociceptive afferents, for example, may initiate startle reactions or other responses that do not require the activation of nociceptors. Furthermore, learned escape responses have been used to test the effect of cortical lesions. This introduces the possibility that the observed deficits may be partly due to a loss of the ability to use the specific sensory information to rapidly develop and execute a learned motor response. Overall, however, most of the evidence supports the interpretation that neurons within somatosensory cortex are necessary for the normal execution of responses to noxious stimuli.

The results of experiments on cortical function in pain mechanisms resemble those found at other supraspinal levels in two important respects. First, there is evidence that structures outside the classic somatosensory and motor pathways, such as limbic cortex, are necessary for normal pain responses and, presumably, pain experience. Second, there is evidence for neural mechanisms that normally attenuate nocifensive responses without interfering with other aspects of motor function.

Conclusion

The available evidence indicates that, at all supraspinal levels, neurons that respond to noxious stimuli may mediate a variety of functions. Some neurons encode discriminative features of the stimulus. Other cells may mediate autonomic or somatomotor behaviors that are characteristic of nocifensive responses. Still others appear to subserve the elaboration of the motivational and affective dimensions of pain. Furthermore the results of our lesion studies in the cat support the accumulating evidence that there are neurons that normally attenuate the nocifensive responses that signal pain.

No single line of experimental evidence is sufficient to determine if a population of supraspinal neurons mediates pain or some critical component of the pain experience. The analysis of supraspinal pain mechanisms in the cat and other mammals requires careful interpretation of the results from a variety of experiments, each designed to contribute to one of the lines of evidence presented in this chapter.

Our studies were supported by National Institutes of Health Grants NS 12015 and NS 12581 and The Veterans Administration.

REFERENCES

1. Albe-Fessard, D., and L. Kruger. Duality of unit discharges from cat centrum medianum in response to natural and electrical stimulation. *J. Neurophysiol.* 25: 3–20, 1962.
2. Angaut-Petit, D. The dorsal column system. II. Functional properties and bulbar relay of the postsynaptic fibres of the cat's fasciculus gracilus. *Exp. Brain Res.* 22: 471–493, 1975.
3. Basbaum, A. I., C. H. Clanton, and H. L. Fields. Three bulbospinal pathways from the rostral medulla of the cat: an autoradiographic study of pain modulating systems. *J. Comp. Neurol.* 178: 209–224, 1978.
4. Basbaum, A. I., and H. L. Fields. Endogenous pain control mechanisms: review and hypothesis. *Ann. Neurol.* 4: 451–462, 1978.
5. Beck, P. W., H. O. Handwerker, and M. Zimmermann. Nervous outflow from the cat's foot during noxious radiant heat stimulation. *Brain Res.* 67: 373–386, 1974.
6. Beitel, R. E., and R. Dubner. Response of unmyelinated (C) polymodal nociceptors to thermal stimuli applied to monkey's face. *J. Neurophysiol.* 39: 1160–1175, 1976.
7. Benton, R. G., and R. B. Mefferd, Jr. The differentiation of escape and avoidance behavior in two thalamic areas in the cat. *Brain Res.* 6: 679–685, 1967.
8. Berkeley, K. J. Response properties of cells in ventrobasal and posterior group nuclei in the cat. *J. Neurophysiol.* 36: 940–952, 1973.
9. Berkeley, K. J. Spatial relationships between the terminations of somatic sensory and motor pathways in the rostral brainstem of cats and monkeys. I. Ascending somatic sensory inputs to lateral diencephalon. *J. Comp. Neurol.* 193: 283–317, 1980.
10. Berkeley, K. J., and R. Parmer. Somatosensory cortical involvement in responses to noxious stimulation in the cat. *Exp. Brain Res.* 20: 363–374, 1974.
11. Bessou, P., and E. R. Perl. Response of cutaneous sensory units with unmyelinated fibers to noxious stimuli. *J. Neurophysiol.* 32: 1025–1043, 1969.
12. Boivie, J. The termination of the cervicothalamic tract in the cat. An experimental study with silver impregnation methods. *Brain Res.* 19: 333–360, 1970.
13. Boivie, J. The termination of the spinothalamic tract in the cat. An experimental study with silver impregnation methods. *Exp. Brain Res.* 12: 331–353, 1971.
14. Brinkhus, H. B., E. Carstens, and M. Zimmermann. Encoding of graded noxious skin heating by neurons in posterior thalamus and adjacent areas in the cat. *Neurosci. Lett.* 15: 37–42, 1979.
15. Brown, A. G., and D. N. Franz. Responses of spinocervical tract neurones to natural stimuli of identified cutaneous receptors. *Exp. Brain Res.* 1: 231–249, 1969.
16. Burgess, P. R., and E. R. Perl. Cutaneous mechanoreceptors and nociceptors. In: *Handbook of Sensory Physiology. Somatosensory System*, edited by A. Iggo. Berlin: Springer-Verlag, 1973, vol. II, p. 59–69.

17. Carmon, A., Y. Friedman, R. Coger, and B. Kenton. Single trial analysis of evoked potentials to noxious thermal stimulation in man. *Pain* 8: 21–32, 1980.
18. Carmon, A., J. Mor, and J. Goldberg. Evoked cerebral responses to noxious thermal stimulation in humans. *Exp. Brain Res.* 25: 103–107, 1976.
19. Carr, K. D., and E. C. Coons. Lateral hypothalamic stimulation gates nucleus gigantocellularis-induced aversion via a reward-independent process. *Brain Res.* 232: 293–316, 1982.
20. Carreras, M., and S. A. Andersson. Functional properties of neurons of the anterior ectosylvian gyrus of the cat. *J. Neurophysiol.* 26: 100–126, 1963.
21. Casey, K. L. Unit analysis of nociceptive mechanisms in the thalamus of the awake squirrel monkey. *J. Neurophysiol.* 29: 727–750, 1966.
22. Casey, K. L. Somatic stimuli, spinal pathways, and size of cutaneous fibers influencing unit activity in the medial medullary reticular formation. *Exp. Neurol.* 25: 35–56, 1969.
23. Casey, K. L. Responses of bulboreticular units to somatic stimuli eliciting escape behavior in the cat. *Int. J. Neurosci.* 2: 15–28, 1971.
24. Casey, K. L. Escape elicited by bulboreticular stimulation in the cat. *Int. J. Neurosci.* 2: 29–34, 1971.
25. Casey, K. L. Neural mechanisms of pain. In: *Handbook of Perception, Feeling and Hurting*, edited by E. C. Carterette and M. P. Friedman. New York: Academic, 1978, vol. VIB, p. 183–230.
26. Casey, K. L. The reticular formation and pain: toward a unifying concept. In: *Pain. Research Publications: Association for Research in Nervous and Mental Disease*, edited by J. J. Bonica. New York: Raven, 1980, vol. 58, p. 93–105.
27. Casey, K. L., G. P. Frommer, T. J. Morrow, and R. G. Voss. Responses of cats to thermal pulses. *Soc. Neurosci. Abstr.* 3: 478, 1977.
27a. Casey, K. L., and T. J. Morrow. Ventral posterior thalamic neurons differentially responsive to noxious stimulation of the awake monkey. *Science.* In press.
28. Chatrian, G. E., R. C. Canfield, T. A. Knauss, and E. Lettich. Cerebral responses to electrical tooth pulp stimulation in man. *Neurology* 25: 745–757, 1975.
29. Craig, A. D., Jr., and H. Burton. Spinal medullary lamina I projection to nucleus submedius in medial thalamus: a possible pain center. *J. Neurophysiol.* 45: 443–466, 1981.
30. Craig, A. D., Jr., S. J. Wiegand, and J. L. Price. The thalamo-cortical projection of the nucleus submedius in the cat. *J. Comp. Neurol.* 206: 28–48, 1982.
31. Curry, M. J. The exteroceptive properties of neurones in the somatic part of the posterior group (PO). *Brain Res.* 44: 439–462, 1972.
32. Dong, W. K., H. Ryu, and I. H. Wagman. Nociceptive responses of neurons in medial thalamus and their relationship to spinothalamic pathways. *J. Neurophysiol.* 41: 1592–1613, 1979.
33. Durinyan, R. A., V. K. Reshetnyak, and A. G. Rabin. Effects of stimulation of somatic cortical areas in unrestrained cats. *Zh. Vyssh. Nervn. Deyat. im. I. P. Pavlova* 22: 109–115, 1972.
34. Foltz, E. L., and L. E. White. Pain "relief" by frontal cingulumotomy. *J. Neurosurg.* 19: 89–100, 1962.
35. Frommer, G. P., B. R. Trefz, and K. L. Casey. Somatosensory function and cortical unit activity in cats with only dorsal column fibers. *Exp. Brain Res.* 27: 113–129, 1977.
36. Gaze, R. M., and G. Gordon. The representation of cutaneous sense in the thalamus of the cat and monkey. *Q. J. Exp. Physiol.* 39: 279–304, 1954.
37. Gebhart, G. F. Opiate and opioid peptide effects on brain stem neurons: relevance to nociception and antinociceptive mechanisms. *Pain* 12: 93–140, 1982.

38. Georgopoulos, A. P. Functional properties of primary afferent units probably related to pain mechanisms in primate glabrous skin. *J. Neurophysiol.* 39: 71–83, 1976.
39. Glassman, R. B. Cutaneous discrimination and motor control following somatosensory cortical ablations. *Physiol. Behav.* 5: 1009–1019, 1970.
40. Glassman, R. B., M. W. Forgus, J. E. Goodman, and H. N. Glassman. Somesthetic effects of damage to cats' ventrobasal complex, medial lemniscus or posterior group. *Exp. Neurol.* 48: 460–492, 1975.
41. Goldman, P. L., W. F. Collins, A. Taub, and J. Fitzmartin. Evoked bulbar reticular unit activity following delta fiber stimulation of somatosensory nerve in cat. *Exp. Neurol.* 37: 597–606, 1972.
42. Graybiel, A. M. The thalamo-cortical projection of the so-called posterior nuclear group: a study with anterograde degeneration methods in the cat. *Brain Res.* 49: 229–244, 1973.
43. Guilbaud, G., J. M. Besson, J. L. Oliveras, and M. C. Wyon-Maillard. Modifications of the firing rate of bulbar reticular units (nucleus gigantocellularis) after intra-arterial injection of bradykinin into the limbs. *Brain Res.* 63: 131–140, 1973.
44. Guilbaud, G., D. Caille, J. M. Besson, and G. Benelli. Single unit activities in ventral posterior and posterior group thalamic nuclei during nociceptive and non-nociceptive stimulations in the cat. *Arch. Ital. Biol.* 115: 38–56, 1977.
45. Halliday, A. M., and V. Logue. Painful sensations evoked by electrical stimulation in the thalamus. In: *Neurophysiology Studied in Man*, edited by G. G. Somjen. Amsterdam: Excerpta Med., 1972, p. 221–230.
46. Halpern, B. P., and J. D. Halverson. Modification of escape from noxious stimuli after bulbar reticular formation lesions. *Behav. Biol.* 11: 215–229, 1974.
47. Harris, F. A. Wide-field neurons in somatosensory thalamus of domestic cats under barbiturate anesthesia. *Exp. Neurol.* 68: 27–49, 1980.
48. Heath, C. J. Distribution of axonal degeneration following lesions of the posterior group of thalamic nuclei in the cat. *Brain Res.* 21: 354–357, 1970.
49. Heath, C. J., and E. G. Jones. An experimental study of ascending connections from the posterior group of thalamic nuclei in the cat. *J. Comp. Neurol.* 141: 397–426, 1971.
50. Henry, J. L., and F. R. Calaresu. Excitatory and inhibitory inputs from medullary nuclei projecting to spinal cardioacceleratory neurons in the cat. *Exp. Brain Res.* 20: 485–504, 1974.
50a. Honda, C. N., S. Mense, and E. R. Perl. Neurons in ventrobasal region of cat thalamus selectively responsive to noxious mechanical stimulation. *J. Neurophysiol.* 49: 662–673, 1983.
51. Jones, E. G., and R. Y. Leavitt. Demonstration of thalamocortical connectivity in the cat somatosensory system by retrograde axonal transport of horseradish peroxidase. *Brain Res.* 63: 414–418, 1973.
52. Jones, E. G., and R. Y. Leavitt. Retrograde axonal transport and the demonstration of non-specific projections to the cerebral cortex and striatum from thalamic intralaminar nuclei in the rat, cat and monkey. *J. Comp. Neurol.* 154: 349–377, 1974.
53. Kaelber, W. W. Subthalamic nociceptive stimulation in the cat: effect of secondary lesion and rostral fiber projections. *Exp. Neurol.* 56: 574–597, 1977.
54. Kaelber, W. W. Escape from and avoidance of nociception elicited by intracranial stimulation of the cat subthalamus. *Exp. Neurol.* 73: 397–420, 1981.
55. Kaelber, W. W., and C. L. Mitchell. The centrum medianum-central tegmental fasciculus complex. A stimulation, lesion and degeneration study in the cat. *Brain* 90: 83–100, 1967.
56. Kaelber, W. W., and C. L. Mitchell. Alteration in escape responding in the cat. A

lesion and degeneration comparison following stimulation studies. *Brain Behav. Evol.* 12: 137–150, 1975.

57. Kaelber, W. W., C. L. Mitchell, A. J. Yarmat, A. K. Afifi, and S. A. Lorens. Centrum medianum-parafascicularis lesions and reactivity to noxious and non-noxious stimulus. *Exp. Neurol.* 46: 282–290, 1975.
58. Kaelber, W. W., and T. B. Smith. Projections of the zona incerta in the cat, with stimulation controls. *Exp. Neurol.* 63: 177–200, 1979.
59. Keene, J. J. Reward-associated inhibition and pain-associated excitation lasting seconds in single intralaminar thalamic units. *Brain Res.* 64: 211–224, 1973.
60. Keene, J. J., and K. L. Casey. Excitatory connection from lateral hypothalamic self-stimulation sites to escape sites in medullary reticular formation. *Exp. Neurol.* 28: 155–166, 1970.
61. Kennard, M. A. The course of ascending fibers in the spinal cord of the cat essential to the recognition of painful stimuli. *J. Comp. Neurol.* 100: 511–524, 1954.
62. Kenshalo, D. R., D. G. Duncan, and C. Weymark. Thresholds for thermal stimulation of the inner thigh, footpad, and face of cats. *J. Comp. Physiol. Psychol.* 63: 133–138, 1967.
63. Kenshalo, D. R., Jr., G. J. Giesler, Jr., R. B. Leonard, and W. D. Willis. Responses of neurons in primate ventral posterior lateral nucleus to noxious stimuli. *J. Neurophysiol.* 43: 1594–1614, 1980.
64. Kerr, F. W. L., and K. L. Casey. Pain, *Neurosci. Res. Program Bull.* 16: 1–207, 1978.

64a. Kniffki, K.-D., and K. Mizumura. Responses of neurons in VPL and VPL-VL region of the cat to algesic stimulation of muscle and tendon. *J. Neurophysiol.* 49: 649–661, 1983.

65. Krupp, P., and M. Monnier. The unspecific intralaminary modulating system of the thalamus. *Int. Rev. Neurobiol.* 9: 45–97, 1966.
66. LeBars, D., A. H. Dickenson, and J. M. Besson. Diffuse noxious inhibitory controls (DNIC). I. Effects on dorsal horn convergent neurones in the rat. *Pain* 6: 283–304, 1979.
67. LeBlanc, H. J., and G. B. Gatipon. Medial bulboreticular response to peripherally applied noxious stimuli. *Exp. Neurol.* 42: 264–273, 1974.
68. Marshall, J. Sensory disturbances in cortical wounds with special reference to pain. *J. Neurol. Psychiatry* 14: 187–204, 1951.
69. Martin, R. F., L. H. Haber, and W. D. Willis. Primary afferent depolarization of identified cutaneous fibers following stimulation in medial brain stem. *J. Neurophysiol.* 42: 779–790, 1979.
70. Mayer, D. J., and D. D. Price. Central nervous system mechanisms of analgesia. *Pain* 2: 379–404, 1976.
71. Mehler, W. R. Some observations on secondary ascending afferent systems in the central nervous system. In: *Pain*, edited by R. S. Knighton and P. R. Dumke. Boston, MA: Little, Brown, 1966, p. 11–32.
72. Mitchell, C. L., and W. W. Kaelber. Effect of medial thalamic lesions on responses elicited by tooth pulp stimulation. *Am. J. Physiol.* 210: 263–269, 1966.
73. Mitchell, C. L., and W. W. Kaelber. Unilateral vs. bilateral medial thalamic lesions and reactivity to noxious stimuli. *Arch. Neurol.* 17: 653–660, 1967.
74. Morrow, T. J., and K. L. Casey. Analgesia produced by mesencephalic stimulation: effect on bulboreticular neurons. In: *Advances in Pain Research and Therapy*, edited by J. J. Bonica and D. Albe-Fessard. New York: Raven, 1976, vol. 1, p. 503–510.
75. Morrow, T. J., and K. L. Casey. A contact thermal stimulator for neurobehavioral research on temperature sensation. *Brain Res. Bull.* 6: 281–284, 1981.
76. Nauta, W. J. H., and H. G. J. M. Kuypers. Some ascending pathways in the brain

stem reticular formation. In: *Reticular Formation of the Brain*, edited by H. H. Jasper and L. D. Proctor. Boston, MA: Little, Brown, 1958, p. 3–30.
77. Niimi, K., M. Niimi, and Y. Okada. Thalamic afferents to the limbic cortex in the cat studied with the method of retrograde axonal transport of horseradish peroxidase. *Brain Res.* 145: 225–238, 1978.
78. Nyberg-Hansen, R. Sites and mode of termination of reticulospinal fibers in the cat. An experimental study with silver impregnation methods. *J. Comp. Neurol.* 124: 71–100, 1965.
79. Nyquist, J. K. Somatosensory properties of neurons of thalamic nucleus ventralis lateralis. *Exp. Neurol.* 48: 123–135, 1975.
80. Nyquist, J. K., and J. H. Greenhoot. Responses evoked from the thalamic centrum medianum by painful input: suppression by dorsal funiculus conditioning. *Exp. Neurol.* 39: 215–222, 1973.
81. Nyquist, J. K., and J. H. Greenhoot. Unit analysis of nonspecific thalamic responses to high-intensity cutaneous input in the cat. *Exp. Neurol.* 42: 609–622, 1974.
82. Oliveras, J. L., F. Redjemi, G. Guilbaud, and J. M. Besson. Analgesia induced by electrical stimulation of the inferior centralis nucleus of the raphe in the cat. *Pain* 1: 139–146, 1975.
83. Ottersen, O. P., and Y. Benari. Afferent connections to the amygdaloid complex of the rat and cat. I. Projections from the thalamus. *J. Comp. Neurol.* 187: 401–424, 1979.
84. Pearl, G. S., and K. V. Anderson. Response patterns of cells in the feline caudal nucleus reticularis gigantocellularis after noxious trigeminal and spinal stimulation. *Exp. Neurol.* 58: 231–241, 1978.
85. Pearl, G. S., and K. V. Anderson. Response of cells in feline nucleus centrum medianum to tooth-pulp stimulation. *Brain. Res. Bull.* 5: 41–45, 1980.
86. Peschanski, M., G. Guilbaud, and M. Gautron. Intralaminar region in rat: neuronal responses to noxious and non-noxious cutaneous stimuli. *J. Exp. Neurol.* 72: 226–238, 1981.
87. Poggio, G. F., and V. B. Mountcastle. A study of the functional contributions of the lemniscal and spinothalamic systems to somatic sensibility. *Bull. Johns Hopkins Hosp.* 108: 266–316, 1960.
88. Price, D. D., and R. Dubner. Neurons that subserve the sensory-discriminative aspects of pain. *Pain* 3: 307–338, 1977.
89. Randa, D. C. Effects of lesions in the cingulate gyrus on the response to tooth pulp stimulation in the cat. *Exp. Neurol.* 30: 423–430, 1971.
90. Rice, C. E., and D. R. Kenshalo. Nociceptive threshold measurements in the cat. *J. Appl. Physiol.* 17: 1009–1012, 1962.
91. Richardson, D. E., and H. Akil. Pain reduction by electrical brain stimulation in man. Pt. I. Acute adminstration in periaqueductal and periventricular sites. *J. Neurosurg.* 47: 178–183, 1977.
92. Robertson, R. T., and S. S. Kaitz. Thalamic connections with limbic cortex. I. Thalamocortical projections. *J. Comp. Neurol.* 195: 501–526, 1981.
93. Sherrington, C. *The Integrative Action of the Nervous System*. New Haven, CT: Yale Univ. Press, 1947, p. 324–330.
94. Siegel, J. M. Behavioral functions of the reticular formation. *Brain Res. Rev.* 1: 69–105, 1979.
95. Strick, P. L. Light microscopic analysis of the cortical projection of the thalamic ventrolateral nucleus in the cat. *Brain Res.* 55: 1–24, 1973.
96. Sumino, R., R. Dubner, and S. Starkman. Responses of small myelinated "warm" fibers to noxious heat stimuli applied to the monkey's face. *Brain Res.* 62: 260–263, 1973.

97. Sweet, W. H. Treatment of medically intractable mental disease by limited frontal leucotomy—justifiable? *N. Engl. J. Med.* 289: 1117–1125, 1973.
98. Whitsel, B. L., L. M. Petrucelli, and G. Werner. Symmetry and connectivity in the map of the body surface in somatosensory area II of primates. *J. Neurophysiol.* 32: 170–183, 1969.
99. Woolsey, C. N., T. C. Erickson, and W. E. Gilson. Localization in somatic sensory and motor areas of human cerebral cortex as determined by direct recording of evoked potentials and electrical stimulation. *J. Neurosurg.* 51: 476–506, 1979.
100. Yaksh, T. L., and D. L. Hammond. Peripheral and central substrates involved in the rostrad transmission of nociceptive information. *Pain* 13: 1–85, 1982.

FIVE

Descending Control of Spinal Nociceptive Transmission

E. Carstens
Department of Animal Physiology,
University of California, Davis, California

It is well known that the human experience of pain varies dramatically, depending on such factors as the situation, emotional state, voluntary control, or therapeutic intervention. There is much recent experimental evidence for the existence of endogenous neural systems that can modulate the central transmission of nociceptive signals in animals. Particularly electrical stimulation or opiate microinjection at sites of the medial brain stem produces analgesia, which is generally measured as an increase in threshold for nocifensive reflexes or behavior (41, 114). Stimulation at these analgesic sites is associated with a powerful descending inhibition of the responses of spinal dorsal horn neurons to noxious skin inputs (88). Many spinal dorsal horn neurons are differentially activated by noxious skin stimuli, and much evidence implicates them in the central transmission of nociceptive information (93). Thus descending inhibition of spinal dorsal horn neurons may be an important mechanism underlying the analgesic effect of central stimulation (82). This chapter focuses on descending modulation of spinal dorsal horn neuronal responses to somatosensory stimuli, with particular emphasis on *1*) recently described brain stem and forebrain regions that cause descending inhibition, *2*) possible functional pathways, *3*) neurotransmitters, and *4*) synaptic mechanisms involved in descending inhibition. The modulation of nociceptive information at supraspinal levels is not considered, although this may also be an important mechanism of analgesia requiring further study.

Medulla and Pons

Electrical stimulation of the medullary raphe nuclei, particularly nucleus raphe magnus (NRM), produces analgesia that is apparently unaccompanied by motor impairment or changes in the animal's responsiveness to nonnoxious stimuli (91). Stimulation of the NRM powerfully inhibits the responses of neurons in laminae I-II and IV-VI of the spinal dorsal horn and of trigeminal nucleus caudalis to noxious mechanical or thermal skin stimuli (11, 36, 42, 46, 53, 78, 84, 102, 107, 113). Inhibition was observed almost immediately (within 13 ms) after onset of NRM stimulation and did not outlast the stimulation period by more than a few hundred milliseconds (113). Excitatory or mixed effects of NRM stimulation were also noted (11, 84). Both wide-dynamic-range (class 2 or lamina V–type) and nociceptive-specific (class 3) neurons, including those projecting in ascending pathways such as the spinothalamic tract (46, 84, 113), were affected by NRM stimulation. Several groups have reported that nonnociceptive (class 1 or lamina IV–type) as well as wide-dynamic-range neuronal responses to nonnoxious skin stimuli were relatively unaffected by NRM stimulation (33, 36, 48, 53, 78). However, others have reported that NRM stimulation strongly inhibited the responses of such neurons to both noxious and nonnoxious skin stimuli (34, 46). The question therefore remains whether the raphe-spinal system selectively modulates nociceptive inputs.

Raphe-spinal inhibition appears to be mediated by a pathway originating from neurons in the raphe nuclei whose axons descend in the dorsolateral funiculi to terminate primarily in laminae I-II and V of the spinal dorsal horn (6, 7, 73, 75, 80). The spinal inhibitory effect of NRM stimulation is blocked by lesions of the dorsolateral funiculi (42, 84, 113). Many raphe-spinal neurons contain the putative neurotransmitter serotonin (5-hydroxytryptamine; 5-HT) (12), and there is considerable evidence for 5-HT involvement in the analgesic and descending inhibitory effects of NRM stimulation. Virtually all 5-HT–containing synaptic terminals in the spinal cord originate from neurons in the lower brain stem (32, 89); 5-HT has a predominantly depressant effect when applied iontophoretically to spinal dorsal horn neurons (11, 62, 70, 96). Descending spinal inhibition produced by NRM stimulation was blocked by depletion of central 5-HT levels with the 5-HT synthesis inhibitor *p*-chlorophenylalanine (PCPA) (102). However, 5-HT antagonists generally failed to reduce descending inhibition produced by NRM stimulation (11, 50, 115), as discussed under **Midbrain**. Endogenous opioid peptides have also been localized within NRM neuronal populations (43), which additionally contain 5-HT and/or project to the spinal cord (66, 108). Both the analgesic and descending spinal inhibitory effects of NRM stimulation in the rat were partially blocked by the opiate antagonist naloxone (90, 101, 119), whereas naloxone did not affect spinal inhibition from NRM stimulation in the cat or monkey

(36, 115). Therefore current evidence does not strongly support a role for opioid peptides in raphe-spinal inhibition but rather more strongly supports such a role for 5-HT.

Bilateral stimulation of the nucleus reticularis gigantocellularis (RGC) and more ventral nucleus reticularis magnocellularis (RMC) of the medullary reticular formation had inhibitory or excitatory effects on spinothalamic neuronal responses to noxious and nonnoxious skin stimuli (55, 83, 84). Inhibition of neuronal responses to sustained cutaneous pressure stimuli was more powerful than to transient cutaneous pressure stimuli (84). Stimulation of the RMC was also reported to produce analgesia that was blocked by the adrenergic antagonist phenoxybenzamine (105), suggesting a role for the putative neurotransmitter norepinephrine in its mediation. Descending projections from neurons in RMC terminate in dorsal and ventral horns, whereas projections from neurons in RGC pass in the ventral spinal white matter to terminate in the ventral horn (6, 7, 73, 75).

Stimulation of the locus ceruleus (LC) in the dorsolateral pons produces analgesia (104) and has a predominantly inhibitory effect on the responses of spinal dorsal horn neurons to noxious and nonnoxious skin stimuli (65). Spinal projections from LC and the subjacent nucleus subceruleus descend in the dorsal and ventral spinal white matter (7, 27, 73, 86). Norepinephrine-containing synaptic terminals in the spinal dorsal horn appear to originate mainly from neurons in LC and other lower brain stem areas (32); norepinephrine, like 5-HT, has a predominantly depressant action on dorsal horn neurons (11, 62). Thus norepinephrine may be involved in the mediation of descending spinal inhibitory effects from LC and other brain stem catecholamine nuclei, although this requires further study.

Recent studies of the synaptic mechanism of raphe-spinal inhibition provide evidence for both presynaptic and postsynaptic actions. Stimulation of the NRM evoked inhibitory postsynaptic potentials (IPSPs) in intracellularly recorded spinothalamic neurons (48), indicating postsynaptic inhibitory action. On the other hand NRM stimulation evoked spinal dorsal root potentials (95) and produced excitability changes at the intraspinal terminals of primary afferent fibers (64, 79), indicating presynaptic inhibitory action. The terminal excitability of C-fiber afferents (including those with skin nociceptors) was reduced by NRM stimulation (64), in contrast to the excitability increase produced by NRM and RGC stimulation in A-fiber afferents (79), which are thought to reflect primary afferent depolarization. Excitability reductions at C-fiber terminals were also produced by 5-HT (18), norepinephrine (69), and opiates (23). Because each of these substances generally has a depressant action on dorsal horn neurons, the excitability reduction at C-fiber terminals may reflect a presynaptic inhibitory action. The possible mechanisms of this action have been discussed elsewhere (18, 64).

Midbrain

Recent work on central pain-modulating systems began with the observation that electrical stimulation of the midbrain periaqueductal gray (PAG) produced analgesia without other noticible sensory or motor impairment (81, 88, 97). Stimulation at analgesic PAG sites powerfully suppressed the responses of spinal dorsal horn neurons to noxious pinching or burning of skin (88). In our electrophysiological studies, we recorded in anesthetized cats the responses of single lumbar dorsal horn units to controlled noxious radiant skin heat stimuli in an attempt to quantitatively analyze descending effects on spinal nociceptive transmission (Fig. 1). Noxious heat stimulation applied to glabrous footpad skin activates a population of nociceptors

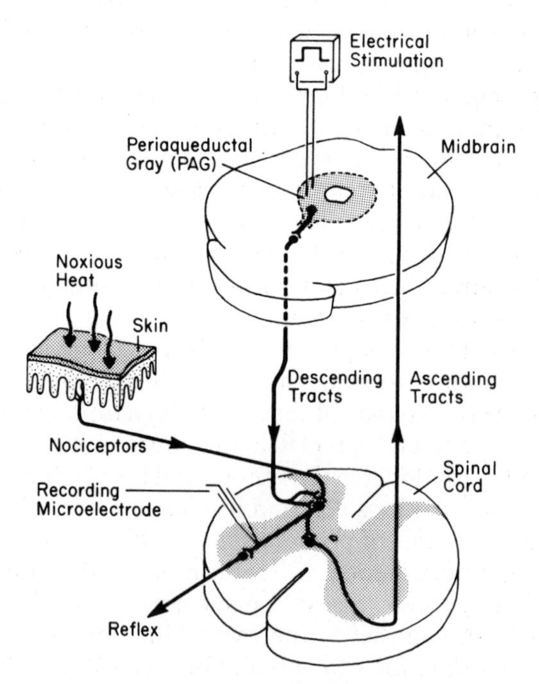

Fig. 1. Experimental setup. Responses of single lumbar dorsal horn units to noxious radiant heat applied to glabrous footpad skin were recorded with microelectrodes in anesthetized cats. To study possible descending effects on spinal neuronal heat-evoked response, midbrain PAG was electrically stimulated via a stereotaxically positioned bipolar electrode. [Adapted from M. Zimmermann (118, Fig. 9-15*A*).]

with afferent C fibers that contact dorsal horn neurons (9). The responses of wide-dynamic-range and nociceptive-specific neurons in laminae I and IV-VI to heat stimuli (50°C, 10 s) repeated at 3-min intervals were stable in magnitude over time (24), and thus a base-line response level could be established against which to compare the actions of descending systems.

Our initial experiments aimed to systematically map midbrain sites where electrical stimulation (100-ms trains at 100 Hz, 3/s) reduced the magnitude of dorsal horn neuronal responses to noxious heat. A bipolar stimulating electrode was stereotaxically lowered into the midbrain in 1-mm-depth intervals, and successive dorsal horn unit responses to 50°C heat stimuli were recorded without (control) and again during concomitant stimulation at each midbrain site. An example of a dorsal horn unit's control response to the 50°C heat stimulus is shown in the upper trace in Figure 2*A*. The middle and lower traces show that the unit's response to the identical 50°C stimulus was reduced during concomitant stimulation at separate sites in the midbrain PAG and lateral midbrain reticular formation (LRF), respectively. Inhibition was seen within seconds after onset of midbrain stimulation (20, 21). Note the rebound in unit firing immediately after termination of PAG and LRF stimulation. Four parallel tracks through the midbrain were made with the stimulating electrode in this experiment. In Figure 2*B* the degree of inhibition of the unit response (expressed as response magnitude during midbrain stimulation in relation to control response) is plotted against the depth of the stimulating electrode along each of the four tracks. Note that 100% indicates no inhibition, and 0% indicates total inhibition of the heat-evoked response, so that the greater width of the darkened area reflects stronger inhibition at that site. The greatest magnitude of inhibition was generated from the dorsal and ventral PAG and underlying tegmentum as well as from a large region spanning the midbrain reticular formation bilaterally (20).

Parametric and pharmacological analyses were also done to compare inhibitory effects of PAG and LRF stimulation. The responses of dorsal horn units to graded increases in temperature of the noxious heat stimulus generally increase linearly from threshold (40–45°C) to 52°C (58). To investigate midbrain effects on noxious heat intensity-coding functions, the temperature series was repeated during concomitant PAG or LRF stimulation. Figure 3*A* shows how a dorsal horn unit's response to 48°C (*upper trace*) was reduced during concomitant PAG or LRF stimulation (*middle* and *lower traces*, respectively). This unit's responses are plotted against stimulus temperature in Figure 3*B* (●), illustrating the relatively linear temperature-response relationship. During PAG stimulation the slope of the temperature-response curve

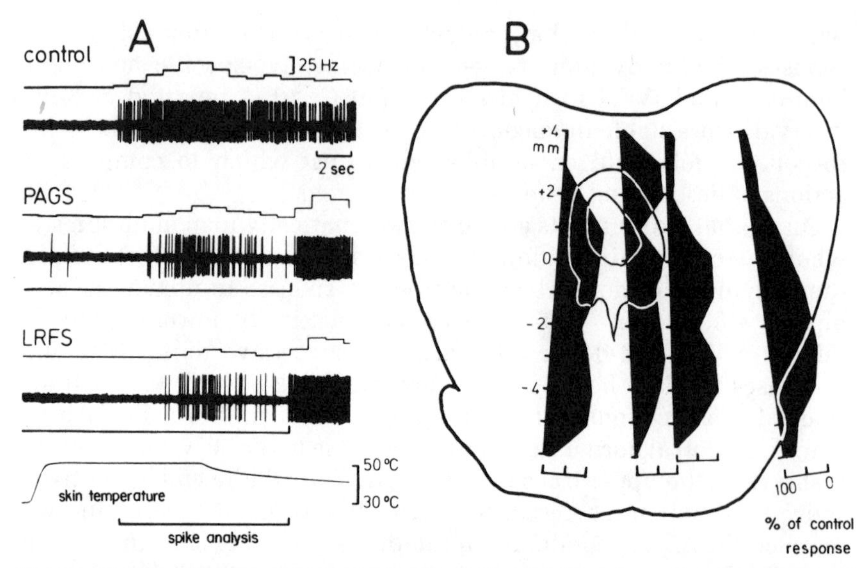

Fig. 2. Map of midbrain sites at which stimulation inhibits responses of spinal dorsal horn unit to noxious skin heating. *A*: oscilloscope traces and peristimulus-time histograms (*top*) of unit's responses to 50°C heat stimuli (time course, *bottom*), without (*upper trace*) and during concomitant electrical stimulation (600 μA) of midbrain periaqueductal gray (*middle trace*, PAGS) or lateral reticular formation (*lower trace*, LRFS). Duration of PAG and LRF stimulation period indicated by *bars* below oscilloscope traces. Responses quantified as total number of impulses during 10-s analysis period (*bottom bar*). *B*: inhibition of dorsal horn unit heat-evoked responses in relation to position of stimulating electrode in midbrain. Successive 50°C skin heat stimuli applied with and without midbrain stimulation (300 μA) at 1-mm-deep intervals along each of 4 penetrations of stimulating electrode (stereotaxic depth coordinates given along left-hand track). Inhibition at each site expressed as magnitude of response during midbrain stimulation in relation to control: 0%, total inhibition; 100%, no inhibition. Right side is ipsilateral to spinal recording site in this and subsequent figures. [From Carstens et al. (20).]

was reduced without any change in the threshold (○). In contrast, during LRF stimulation the temperature-response curve was shifted in a parallel manner to the right with a large increase in threshold (20, 24). Similar differential actions of PAG and LRF stimulation are shown in Figure 3C for another unit. These results suggest that stimulation of PAG and LRF may activate functionally separate descending inhibitory systems that can independently modulate the firing threshold and gain of neuronal responses to noxious heat. This is further supported by pharmacological data. The inhibitory effect of PAG, but not LRF, stimulation was blocked by acute systemic administration of the 5-HT antagonist methysergide (17). In addition, inhibition produced by PAG stimulation was significantly weaker in animals pretreated with PCPA, whereas that produced by LRF stimulation was stronger.

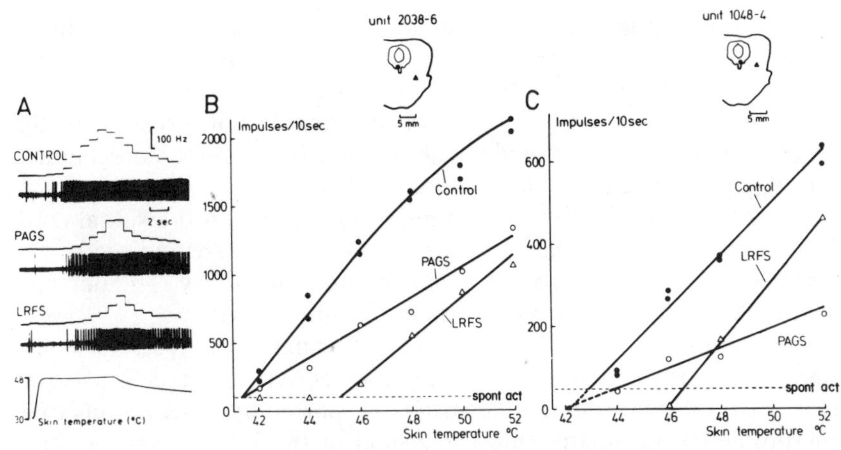

Fig. 3. Intensity coding by dorsal horn units for graded noxious heat stimuli: differential effects of PAG and LRF stimulation. *A*: oscilloscope traces and histograms of unit responses to 48°C heat stimuli, without (*upper*) and during PAG (*middle*) and LRF (*lower*) stimulation. *B*: integrated unit responses against stimulus temperature, without (●) and during PAG (○) and LRF (△) stimulation (450 μA). *Inset*: stimulation sites. *C*: graph as in *B* for another unit. [From Carstens et al. (20).]

The analgesic effect of PAG stimulation was also blocked by PCPA (4) and 5-HT antagonists (59). These results indicate that 5-HT is involved in the mediation of analgesia and descending inhibition produced by PAG but not LRF stimulation. It is not yet known which transmitter substances mediate descending inhibition from the LRF.

As was the case with NRM stimulation, there are discrepancies as to whether PAG stimulation selectively inhibits the responses of spinal dorsal horn neurons to noxious as opposed to nonnoxious skin stimuli (36, 88, 107, 110, 116). We found that the responses to controlled brushing of fur of both nonnociceptive and wide-dynamic-range dorsal horn units were equally reduced (to a mean of 80% of control) by PAG and LRF stimulation, which, however, more powerfully inhibited wide-dynamic-range unit responses to noxious heating (16).

Current evidence indicates that descending inhibition from the PAG may be mediated via activation of the raphe-spinal pathway (41). The PAG projects sparsely to the spinal cord (7, 26, 73, 111) but has a prominent projection to the medullary NRM (2, 45) as well as the RMC (2). Stimulation of the PAG predominantly excites NRM neurons including those with spinal projections (10, 40, 77), and both the analgesic (8) and descending inhibitory effects (E. Carstens, unpublished observations) of PAG stimulation are blocked by lesions of the spinal dorsolateral funiculi in which the raphe-spinal fibers pass. It has been suggested that the PAG-NRM pathway may utilize 5-HT as an excitatory transmitter whose action, when blocked by 5-HT antagonists, would account for the abolition of descending inhibition produced by PAG but not NRM stimulation (17, 50, 115). Apparently, however, 5-HT–containing midbrain neurons do not project directly to NRM (2). The site and mode of action of 5-HT along the descending pathway require further clarification.

The inhibitory effect of LRF stimulation may reach the spinal cord directly via reticulospinal projections (26, 111) and/or indirectly via connections with neurons in LC or the pontomedullary reticular formation (37), which in turn descend to the spinal cord.

Stimulation of the midbrain reticular formation is aversive, commensurate with a role in ascending nociceptive transmission. However, an additional role in analgesia is supported by observations that microinjection of opiates into the region of the LRF, as well as into PAG, produced naloxone-reversible analgesia (114). Analgesia produced by PAG stimulation was partially blocked by naloxone (5), suggesting a role for endogenous opiates in midbrain analgesic mechanisms. On the other hand naloxone did not affect descending spinal inhibition produced by PAG (19, 36, 61, 115) or LRF (25) stimulation, indicating that endogenous opiates do not play an important role under the experimental conditions. The relative contribution of descending inhibitory pathways to mechanisms of analgesia and the role played by endogenous opiates must still be studied.

If descending effects produced by PAG stimulation are mediated via activation of the raphe-spinal pathway, both presynaptic and postsynaptic inhibitory mechanisms may be involved (see **Medulla and Pons**). The possibility of a presynaptic inhibitory action is further supported by our observations that PAG stimulation generally reduced the excitability at the intraspinal terminals of afferent C fibers (E. Carstens, H. Gilly, H. Schreiber, and M. Zimmermann, manuscript in preparation). Stimulation of the LRF occasionally produced increases or decreases but usually had little effect on C-fiber terminal excitability.

The spontaneous and peripherally evoked activity of spinal dorsal horn neurons increases after spinal transection (58, 112), indicating that a tonically active descending inhibitory influence has been removed. The origin of tonic descending inhibition is not known. Tonic inhibition was not affected by NRM lesions (56) or by antagonists of 5-HT or other putative transmitters (35), implicating some pathway other than the raphe-spinal tract. On the other hand the threshold for nocifensive reactions can be reduced by lesions of NRM (94), suggesting that the raphe-spinal tract may tonically inhibit nociceptive transmission. Another interesting observation is that linear intensity-coding functions of dorsal horn units for graded noxious heat stimuli are shifted in a parallel manner to the left (i.e., toward lower temperatures) after spinalization (85, 117), as if a potential tonic inhibitory effect from the LRF system might have been removed. Further studies are needed to evaluate the possible contribution of reticulospinal projections to tonic descending inhibition.

Forebrain

Analgesia is produced by electrical stimulation at a variety of sites including the medial diencephalon (67, 81, 87, 99, 100, 106), lateral hypothalamus (31), ventrobasal thalamus (54), medial preoptic and septal areas (1, 49, 63, 87), caudate nucleus (76), and internal capsule (3). Until recently, however, there was no information as to whether these analgesic effects might be mediated by descending inhibition of spinal nociceptive transmission. The aim of my recent work was to systematically map these and other areas in the forebrain that might influence spinal dorsal horn neurons.

In the initial series of experiments, the stimulating electrode was systematically lowered into the brain along tracks at different anteroposterior (AP) levels through the medial diencephalon to determine sites where stimulation affected the responses of dorsal horn units to 50°C heat stimuli. In the example shown in Figure 4*A*, each vertical pair of peristimulus-time histograms shows a unit's responses to 50°C heat stimuli without (*upper histogram*) and during (*lower histogram*) concomitant brain stimulation, when the stimulating electrode was positioned at the indicated stereotaxic depth coordinate (DV) along a track at AP +7.5. Stimulation at DV 0 abolished and at DV −1 reduced the response, whereas stimulation at DV +3 and DV −3 had little effect. The time course of inhibition was generally restricted to the period of brain stimulation. In Figure 4*B*, graphs of inhibition versus depth along the track of the stimulating electrode are superimposed on histological sections at the four indicated AP levels. The most powerful inhibition was clearly generated in the periventricular gray (PVG) at each level along a region extending from the posterior (+7.5) through the anterior (+12.5) hypothalamic areas. The inhibitory zone is continuous posteriorly with the midbrain PAG. It extends anteriorly into the medial preoptic area, as illustrated in Figure 5. The histograms in Figure 5*A* show how another dorsal horn unit's heat-evoked responses were inhibited by stimulation at DV −2 and DV −3 along a track at AP +14.5. The most powerful inhibition was generated in the medial preoptic area ventral to the anterior commissure (Fig. 5*B*). Little inhibition was generated along another track at AP +16.5. Further experiments revealed that the inhibitory region extends into the ventromedial septal area of the basal forebrain. Figure 6 illustrates

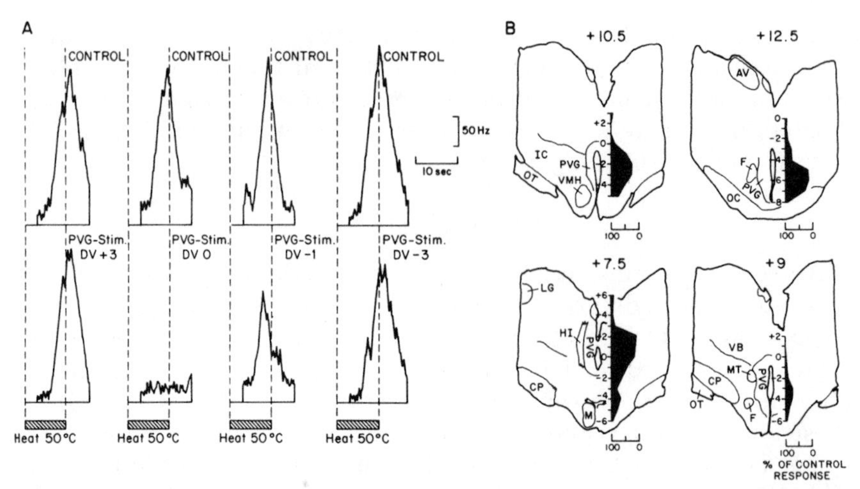

Fig. 4. Map of medial diencephalic sites at which stimulation inhibits responses of spinal dorsal horn unit to noxious skin heating. *A*: peristimulus-time histogram of unit's responses to 50°C heat stimuli, without (controls, *upper row*) and during stimulation (200 μA) at indicated stereotaxic depth coordinates (DV) along a track at anterior level +7.5. Brain stimulation began 10 s prior to onset of heat stimulus and lasted 25 s. *B*: inhibition in relation to depth of stimulating electrode along 4 tracks at indicated anterior levels (format as in Fig. 2*B*). Abbreviations for this and subsequent figures: AC, anterior commissure; AV, anterior ventral n.; CC, corpus callosum; Cd, caudate n.; CP, cerebral peduncle; F, fornix; HI, habenulointerpeduncular tract; IC, internal capsule; LG, lateral geniculate n.; LV, lateral ventricle; M, mamillary body; MS, medial septal n.; MT, mamillothalamic tract; OC, optic chiasm; OT, optic tract; PVG, periventricular gray; VB, ventrobasal thalamic nuclei; VMH, ventromedial hypothalamic n. [From Carstens (15).]

that powerful inhibition was generated ventromedially in the region of the diagonal band at AP +16, whereas only weak inhibition was generated in the medial septal nucleus. Stimulation along another track at AP +17 no longer generated inhibition but rather mild facilitation of unit heat-evoked responses. In summary, descending inhibition can be generated along a medial region extending from the midbrain PAG through the diencephalic PVG and preoptic area into the posterior portion of the basal telencephalon.

Parametric and pharmacological studies were also done to compare the inhibitory effects of medial diencephalic and midbrain PAG stimulation. As was the case with PAG stimulation, stimulation of the PVG (15) and medial preoptic and septal areas (22) reduced the slope of the linear intensity-coding functions of dorsal horn units for graded noxious heat stimuli. An example of such an effect produced by medial preoptic stimulation is shown in Figure 7. In addition, the inhibitory effects of PVG, medial preoptic, and septal stimulation were blocked

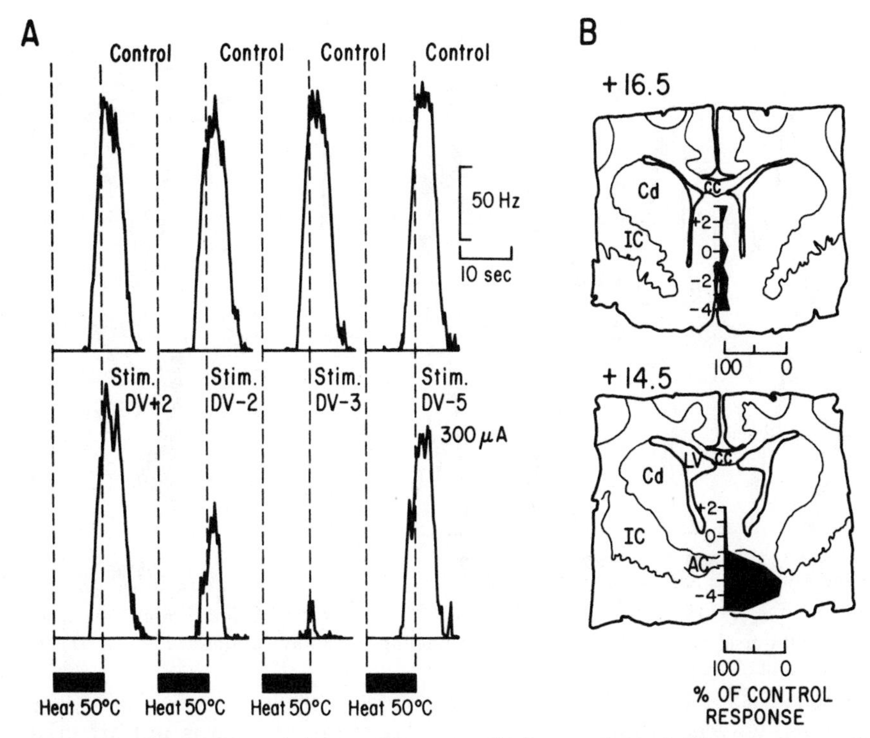

Fig. 5. Map of sites in medial preoptic area at which stimulation inhibits responses of spinal dorsal horn unit to noxious skin heating (for abbreviations see Fig. 4). [From Carstens et al. (22).]

or reduced by methysergide and were significantly weaker in PCPA-pretreated than in untreated animals (E. Carstens, J. D. MacKinnon, and M. J. Guinan, manuscript in preparation). These results point to a role for 5-HT in the mediation of descending inhibition from these areas. Naloxone did not affect descending inhibition from these areas, indicating that endogenous opiates are not involved (E. Carstens, unpublished observations).

Descending inhibition from the PVG could conceivably be mediated via direct hypothalamospinal projections (7, 57, 68, 73). However, 5-HT appears to be involved in the mediation of this inhibition, but 5-HT–containing neurons are not found in the diencephalon or spinal cord (108; cf. 44). A more likely pathway therefore might involve projections from the medial hypothalamus to the midbrain PAG (28, 29, 51) or medullary NRM (2), from which serotonergic descending inhibition is generated. Medial hypothalamic stimulation predominantly excites PAG neurons (103). In behavioral experiments, analgesia produced by medial diencephalic stimulation was blocked by PAG

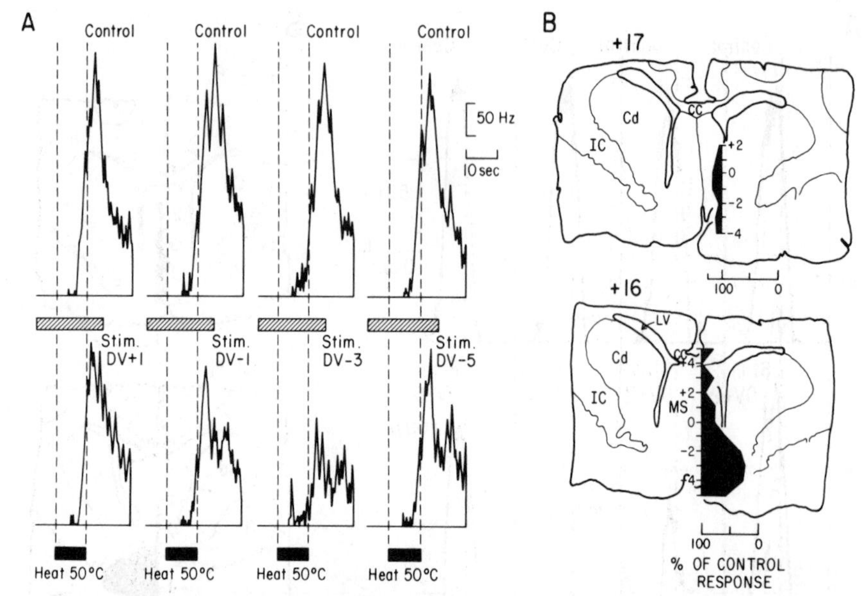

Fig. 6. Map of sites in medial basal forebrain at which stimulation inhibits responses of spinal dorsal horn unit to noxious skin heating (for abbreviations see Fig. 4). [From Carstens et al. (22).]

lesions, whereas that produced by PAG stimulation was not affected by medial diencephalic lesions, indicating the involvement of a pathway descending to or through the PAG (98). Descending inhibition from the ventromedial septal area may involve a serial pathway via connections with the medial preoptic area (72, 109), which in turn connects with more posterior areas of the medial hypothalamus (28). Alternatively, inhibition from the ventromedial septal area may involve a projection to the lateral hypothalamic area (72, 109).

The descending inhibitory effects of midbrain PAG, medial diencephalic, and septal stimulation are parametrically and pharmacologically very similar, suggesting that these areas might constitute portions of a continuous, functionally homogeneous inhibitory system. Further studies were done to determine the lateral extent of this inhibitory region, as well as the possible contribution of other diencephalic structures to descending inhibition (E. Carstens, M. Fraunhoffer, and J. D. MacKinnon, manuscript in preparation). In these experiments a comb of parallel stimulating electrodes spaced at 2-mm intervals was systematically lowered into the brain, and the effects of stimulation at each site on dorsal horn unit responses to 50°C heat stimuli were determined. Figure 8 illustrates results of experiments in which the ipsilateral (*A*) and contralateral (*B*) diencephalon were mapped. Ipsi-

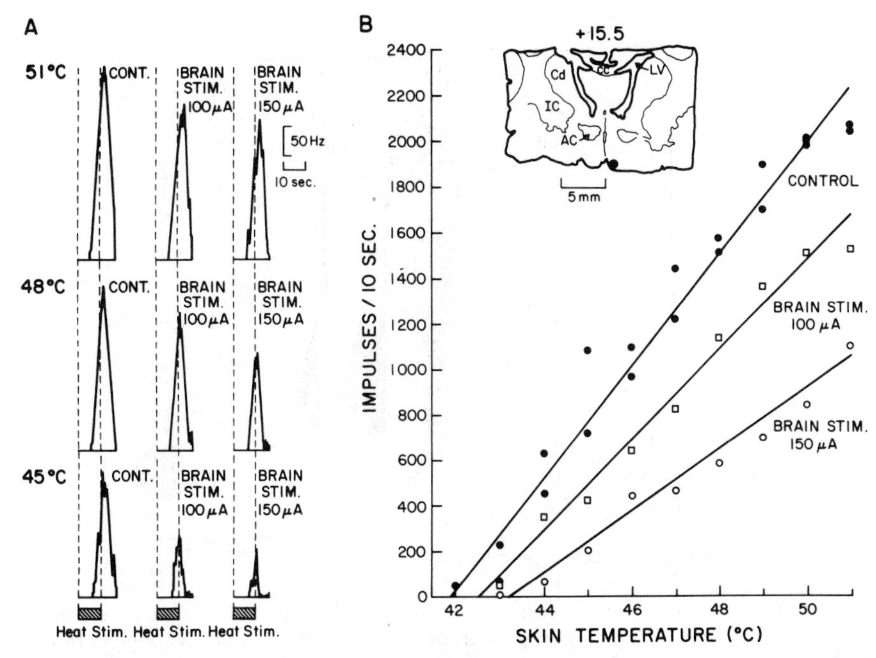

Fig. 7. Effect of medial preoptic stimulation on intensity coding by spinal dorsal horn unit for graded noxious heat stimuli. *A*: histograms of unit responses to heat stimuli at indicated temperatures, without (*left column*) and during medial preoptic stimulation at 100 μA (*middle column*) and 150 μA (*right column*). *B*: same unit's responses against stimulus temperature without (control, ●) and during 100 μA (□) and 150 μA (○) stimulation at site shown in *inset* (for abbreviations see Fig. 4). [From Carstens et al. (22).]

laterally, powerful inhibition was generated in the midline thalamus and hypothalamus. There was little or no inhibition, or mild facilitation, at progressively more lateral sites in the ventrobasal thalamic nuclei. At ventrolateral sites, however, powerful inhibition was generated in a region encompassing the lateral hypothalamic area. Inhibition was also generated in the cerebral peduncle. Contralaterally, inhibition was generated in the medial thalamus and hypothalamus, the lateral hypothalamic area, the cerebral peduncle, and in the ventrobasal thalamic nuclei and optic tract region. It was recently reported that stimulation of the ventral posterior thalamic nucleus inhibited primate spinothalamic neurons (47). Descending inhibition was also generated from the internal capsule bilaterally. The inhibitory effects of peduncular and capsular stimulation suggest the involvement of corticofugal fibers. Stimulation of the sensorimotor cortex and pyramidal tract was previously shown to have inhibitory and excitatory effects on spinal neurons (14, 30, 39, 107, 112), but there is little

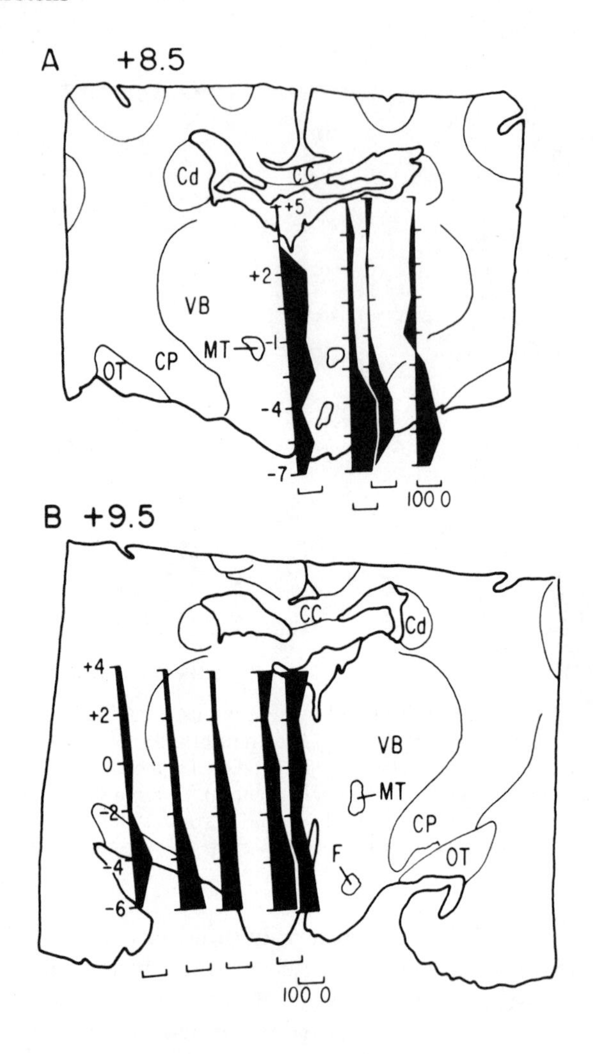

Fig. 8. Map of diencephalic sites at which stimulation inhibits spinal dorsal horn unit responses to noxious skin heating. Tracks made through ipsilateral (*A*) and contralateral (*B*) diencephalon with comb of parallel stimulating electrodes (for abbreviations see Fig. 4). [From E. Carstens, M. Fraunhoffer, and J. D. MacKinnon (manuscript in preparation).]

evidence regarding corticospinal effects on nociceptive transmission. We found that stimulation within the hindlimb region of the contralateral somatosensory cortex, at sites comparable to those from which cat spinocervical neurons were inhibited (14), powerfully inhibited the 50°C heat-evoked responses of most dorsal horn units studied to date (E. Carstens, M. Fraunhoffer, and J. D. MacKinnon, manuscript in preparation).

Parametric and pharmacological data were also obtained on the inhibitory effects produced by stimulation of the lateral hypothalamus, the cerebral cortex and peduncles, and the internal capsule. Lateral hypothalamic stimulation produced slope reductions in the intensity-coding functions of dorsal horn units for graded noxious heat stimuli, and the inhibition was reduced by methysergide, indicating a partial involvement of 5-HT in its mediation (E. Carstens, M. Fraunhoffer, and S. Suberg, manuscript in preparation). This inhibition could be mediated via lateral hypothalamic projections to the midbrain PAG (51, 92) or medullary NRM (2, 92). Similar effects on intensity coding were also produced by stimulation of the cerebral cortex and peduncles and the internal capsule. Methysergide generally did not affect this inhibition, however, indicating involvement of a nonserotonergic descending pathway. The possible role of direct corticospinal projections is currently being investigated.

Conclusions

The complete extent of brain areas contributing to descending modulation of spinal nociceptive transmission is not yet known, but it is already apparent that widespread regions of the brain stem and forebrain are involved. Structures associated with a variety of behavioral functions are included, such as the ascending transmission of somatosensory (including nociceptive) information, autonomic and endocrine control, emotion, sleep, and mechanisms of alerting and attention. This chapter stresses descending inhibitory control of spinal dorsal horn neuronal responses to noxious skin inputs, but the brain regions discussed might play a more general role in descending modulation of all types of spinal somatosensory input during various behavioral states.

There is a striking correspondence between central sites at which stimulation produces analgesia and sites thus far studied at which stimulation powerfully inhibits the responses of spinal dorsal horn neurons to noxious skin inputs. This correspondence strengthens the argument that descending inhibition may be an important mechanism underlying the analgesic effect of central stimulation. However, the inhibitory effect of stimulation at these sites has a rapid time course, whereas stimulation-produced analgesia has been frequently reported to outlast the stimulation period by minutes or hours (81, 87). Thus stimulation-produced analgesia may be mediated by neural processes of longer time course in addition to phasically active descending inhibitory pathways.

Pain perception in humans and painlike behavior in animals can be affected by a wide variety of factors, such as opiate drugs, acupuncture, peripheral stimulation, emotional state, stress, voluntary control, and hypnosis. The analgesic effects of some of these factors may be mediated via anatomically or pharmacologically separate mechanisms, as has been shown for opiate-induced compared to stress-induced analgesia (60). Conceivably, different pain-modulating factors may operate via the selective activation of separate descending control systems in the brain. For example, the proposed serotonergic PAG–raphe-spinal inhibitory system may be activated from an extensive core region extending into the basal forebrain. This system might be capable of modulating the gain of nociceptive transmission by suppressing the responses of spinal dorsal horn neurons to suprathreshold noxious stimuli without altering the neuronal firing threshold. Such a mechanism might be relevant to pain tolerance. Interestingly, medial diencephalic stimulation was reported to reduce human chronic and acute pain without altering the thermal pain threshold (67, 100). Other brain areas might raise the firing threshold of spinal dorsal horn neurons via activation of the nonserotonergic midbrain LRF system. Analgesia produced by stimulation at a number of central sites is frequently characterized by an increased threshold for nocifensive reactions. Finally, the activity of spinal dorsal horn neurons may be independently modulated by other parallel descending pathways (e.g., corticospinal tract), but the details must still be worked out.

The natural means by which descending control systems are activated and their normal functional role remain questions requiring further research. Many neurons in the midbrain PAG and reticular formation respond to nonnoxious and noxious somatosensory stimuli and may receive multimodal inputs (13, 38). If such neurons project into descending systems to inhibit spinal dorsal horn neurons in a negative-feedback loop, then there would be self-limitation in the somatosensory-evoked activity of the spinal neurons. Alternatively, descending control systems might inhibit the background activity of all spinal dorsal horn neurons except those directly activated by a noxious peripheral stimulus, thus serving to reduce the signal-to-noise ratio for ascending nociceptive signals (74). In either case the level of activity in each descending control system involved would be subject to the complex array of inputs coming from other brain areas.

REFERENCES

1. Abbot, F. V., and R. Melzack. Analgesia produced by stimulation of limbic structures and its relation to epileptiform after-discharges. *Exp. Neurol.* 62: 720–734, 1978.

2. Abols, I. A., and A. I. Basbaum. Afferent connections of the rostral medulla of the cat: a neural substrate for midbrain-medullary interactions in the modulation of pain. *J. Comp. Neurol.* 201: 285–297, 1981.
3. Adams, J. E., Y. Hosobuchi, and H. L. Fields. Stimulation of internal capsule for relief of chronic pain. *J. Neurosurg.* 41: 740–744, 1974.
4. Akil, H., and J. C. Liebeskind. Monoaminergic mechanisms of stimulation-produced analgesia. *Brain Res.* 94: 279–296, 1975.
5. Akil, H., D. J. Mayer, and J. C. Liebeskind. Antagonism of stimulation-produced analgesia by naloxone, a narcotic antagonist. *Science* 191: 961–962, 1976.
6. Basbaum, A. I., C. H. Clanton, and H. L. Fields. Three bulbospinal pathways from the rostral medulla of the cat: an autoradiographic study of pain modulating systems. *J. Comp. Neurol.* 178: 209–224, 1978.
7. Basbaum, A. I., and H. L. Fields. The origin of descending pathways in the dorsolateral funiculus of the spinal cord of the cat and rat: further studies on the anatomy of pain modulation. *J. Comp. Neurol.* 187: 513–532, 1979.
8. Basbaum, A. I., N. J. E. Marley, J. O'Keefe, and C. H. Clanton. Reversal of morphine and stimulus-produced analgesia by subtotal spinal cord lesions. *Pain* 3: 43–56, 1977.
9. Beck, P. W., H. O. Handwerker, and M. Zimmermann. Nervous outflow from the cat's foot during noxious radiant heat stimulation. *Brain Res.* 67: 373–386, 1974.
10. Behbehani, M., and H. L. Fields. Evidence that an excitatory connection between the periaqueductal gray and nucleus raphe magnus mediates stimulation produced analgesia. *Brain Res.* 170: 85–93, 1979.
11. Belcher, G., R. W. Ryall, and R. Schaffner. The differential effects of 5-hydroxytryptamine, noradrenaline and raphe stimulation on nociceptive and nonnociceptive dorsal horn interneurones in the cat. *Brain Res.* 151: 307–321, 1978.
12. Bowker, R. M., H. W. M. Steinbusch, and J. D. Coulter. Serotonergic and peptidergic projections to the spinal cord demonstrated by a combined retrograde HRP histochemical and immunocytochemical staining method. *Brain Res.* 211: 412–417, 1981.
13. Bowsher, D. Role of the reticular formation in responses to noxious stimulation. *Pain* 2: 361–378, 1976.
14. Brown, A. G., J. D. Coulter, P. K. Rose, A. D. Short, and P. J. Snow. Inhibition of spinocervical tract discharges from localized areas of the sensorimotor cortex in the cat. *J. Physiol. London* 264: 1–16, 1977.
15. Carstens, E. Inhibition of spinal dorsal horn neuronal responses to noxious skin heating by medial hypothalamic stimulation in the cat. *J. Neurophysiol.* 48: 808–822, 1982.
16. Carstens, E., H. Bihl, D. R. F. Irvine, and M. Zimmermann. Descending inhibition from medial and lateral midbrain of spinal dorsal horn neuronal responses to noxious and nonnoxious cutaneous stimuli in the cat. *J. Neurophysiol.* 45: 1029–1042, 1981.
17. Carstens, E., M. Fraunhoffer, and M. Zimmermann. Serotonergic mediation of descending inhibition from midbrain periaqueductal gray, but not reticular formation, of spinal nociceptive transmission in the cat. *Pain* 10: 149–167, 1981.
18. Carstens, E., D. Klumpp, M. Randic, and M. Zimmermann. Effect of iontophoretically applied 5-hydroxytryptamine on the excitability of single primary afferent C- and A-fibers in the cat spinal cord. *Brain Res.* 220: 151–158, 1981.
19. Carstens, E., D. Klumpp, and M. Zimmermann. The opiate antagonist, naloxone, does not affect descending inhibition from midbrain of nociceptive spinal neuronal discharges in the cat. *Neurosci. Lett.* 11: 323–327, 1979.
20. Carstens, E., D. Klumpp, and M. Zimmermann. Differential inhibitory effects of

medial and lateral midbrain stimulation on spinal neuronal discharges to noxious skin heating in the cat. *J. Neurophysiol.* 43: 332–342, 1980.

21. Carstens, E., D. Klumpp, and M. Zimmermann. Time course and effective sites for inhibition from midbrain periaqueductal gray of spinal dorsal horn neuronal responses to cutaneous stimuli in the cat. *Exp. Brain Res.* 38: 425–430, 1980.
22. Carstens, E., J. D. MacKinnon, and M. J. Guinan. Inhibition of spinal dorsal horn neuronal responses to noxious skin heating by medial preoptic and septal stimulation in the cat. *J. Neurophysiol.* 48: 981–991, 1982.
23. Carstens, E., I. Tulloch, W. Zieglgänsberger, and M. Zimmermann. Presynaptic excitability changes induced by morphine in single cutaneous afferent C- and A-fibers. *Pfluegers Arch.* 379: 143–147, 1979.
24. Carstens, E., T. Yokota, and M. Zimmermann. Inhibition of spinal neuronal responses to noxious skin heating by stimulation of mesencephalic periaqueductal gray in the cat. *J. Neurophysiol.* 42: 558–568, 1979.
25. Carstens, E., and M. Zimmermann. The opiate antagonist naloxone does not consistently block inhibition of spinal nociceptive transmission produced by stimulation in lateral midbrain reticular formation of the cat. *Neurosci. Lett.* 20: 335–339, 1980.
26. Castiglioni, A. J., M. C. Gallaway, and J. D. Coulter. Spinal projections from the midbrain in monkey. *J. Comp. Neurol.* 178: 329–346, 1978.
27. Commissiong, J. W., S. O. Hellstrom, and N. H. Neff. A new projection from locus coeruleus to the spinal ventral columns: histochemical and biochemical evidence. *Brain Res.* 148: 207–213, 1978.
28. Conrad, L. C. A., and D. W. Pfaff. Efferents from medial basal forebrain and hypothalamus in the rat. I. An autoradiographic study of the medial preoptic area. *J. Comp. Neurol.* 169: 185–220, 1976.
29. Conrad, L. C. A., and D. W. Pfaff. Efferents from medial basal forebrain and hypothalamus in the rat. II. An autoradiographic study of the anterior hypothalamus. *J. Comp. Neurol.* 169: 221–262, 1976.
30. Coulter, J. D., R. A. Maunz, and W. D. Willis. Effects of stimulation of sensorimotor cortex on primate spinothalamic neurons. *Brain Res.* 65: 351–356, 1974.
31. Cox, V. C., and E. S. Valenstein. Attenuation of aversive properties of peripheral shock by hypothalamic stimulation. *Science* 149: 323–325, 1965.
32. Dahlstrom, A., and K. Fuxe. Evidence for the existence of monoamine containing neurons in the central nervous system. II. Experimentally induced changes in the intraneuronal amine levels of bulbospinal neuron systems. *Acta Physiol. Scand. Suppl.* 247: 1–36, 1965.
33. Dawson, N. J., A. H. Dickenson, R. F. Hellon, and C. J. Woolf. Inhibitory controls on thermal neurones in the spinal trigeminal nucleus of cats and rats. *Brain Res.* 209: 440–445, 1981.
34. Dostrovsky, J. O. Raphe and periaqueductal gray induced suppression of non-nociceptive neuronal responses in the dorsal column nuclei and trigeminal subnucleus caudalis. *Brain Res.* 200: 184–189, 1980.
35. Duggan, A. W. Pharmacological studies of tonic supraspinal inhibition of dorsal horn neurones excited by noxious cutaneous stimuli. In: *Spinal Cord Sensation*, edited by A. G. Brown and M. Rethelyi. Edinburgh: Scottish Academic, 1981, p. 243–252.
36. Duggan, A. W., and B. T. Griersmith. Inhibition of the spinal transmission of nociceptive information by supraspinal stimulation in the cat. *Pain* 6: 149–161, 1979.
37. Edwards, S. B. Autoradiographic studies of the projections of the midbrain reticular

formation: descending projections of nucleus cuneiformis. *J. Comp. Neurol.* 161: 341–358, 1975.

38. Eickhoff, R., H. O. Handwerker, D. S. McQueen, and E. Schick. Noxious and tactile input to medial structures of midbrain and pons in the rat. *Pain* 5: 99–113, 1978.
39. Fetz, E. E. Pyramidal tract effects on interneurons in the cat lumbar dorsal horn. *J. Neurophysiol.* 31: 69–80, 1968.
40. Fields, H. L., and S. D. Anderson. Evidence that raphe-spinal neurons mediate opiate and midbrain stimulation-produced analgesias. *Pain* 5: 333–349, 1978.
41. Fields, H. L., and A. I. Basbaum. Brainstem control of spinal pain-transmission neurons. *Annu. Rev. Physiol.* 40: 217–248, 1978.
42. Fields, H. L., A. I. Basbaum, C. H. Clanton, and S. D. Anderson. Nucleus raphe magnus inhibition of spinal cord dorsal horn neurons. *Brain Res.* 126: 441–453, 1977.
43. Finley, J. C. W., J. L. Maderdrut, and P. Petrusz. The immunocytochemical localization of enkephalin in the central nervous system of the rat. *J. Comp. Neurol.* 198: 541–565, 1981.
44. Frankfurt, M., J. M. Lauder, and E. C. Azmitia. The immunocytochemical localization of serotonergic neurons in the rat hypothalamus. *Neurosci. Lett.* 24: 227–232, 1981.
45. Gallager, D. W., and A. Pert. Afferents to brain stem nuclei (brain stem raphe, nucleus reticularis pontis caudalis and nucleus gigantocellularis) in the rat as demonstrated by microiontophoretically applied horseradish peroxidase. *Brain Res.* 144: 257–275, 1978.
46. Gerhart, K. D., T. K. Wilcox, J. M. Chung, and W. D. Willis. Inhibition of nociceptive and nonnociceptive responses of primate spinothalamic cells by stimulation in medial brain stem. *J. Neurophysiol.* 45: 121–136, 1981.
47. Gerhart, K. D., R. P. Yezierski, T. K. Wilcox, A. E. Grossman, and W. D. Willis. Inhibition of primate spinothalamic tract neurons by stimulation in ipsilateral or contralateral ventral posterior lateral (VPL_c) thalamic nucleus. *Brain Res.* 229: 514–519, 1981.
48. Giesler, G. J., Jr., K. D. Gerhart, R. P. Yezierski, T. K. Wilcox, and W. D. Willis. Postsynaptic inhibition of primate spinothalamic neurons by stimulation in nucleus raphe magnus. *Brain Res.* 204: 184–188, 1981.
49. Gol, A. Relief of pain by electrical stimulation of the septal area. *J. Neurol. Sci.* 5: 115–120, 1967.
50. Griersmith, B. T., A. W. Duggan, and R. A. North. Methysergide and supraspinal inhibition of the spinal transmission of nociceptive information in the anesthetized cat. *Brain Res.* 204: 147–158, 1981.
51. Grofova, I., O. P. Ottersen, and E. Rinvik. Mesencephalic and diencephalic afferents to the superior colliculus and periaqueductal gray substance demonstrated by retrograde axonal transport of horseradish peroxidase in the cat. *Brain Res.* 146: 205–220, 1978.
52. Guilbaud, G., J. M. Besson, J. L. Oliveras, and J. C. Liebeskind. Suppression by LSD of the inhibitory effect exerted by dorsal raphe stimulation on certain spinal cord interneurons in the cat. *Brain Res.* 61: 417–422, 1973.
53. Guilbaud, G., J. L. Oliveras, G. J. Giesler, Jr., and J. M. Besson. Effects induced by stimulation of the centralis inferior nucleus of the raphe on dorsal horn interneurons in cat's spinal cord. *Brain Res.* 126: 355–360, 1977.
54. Gybels, J., J. van Hees, and F. Peluso. Modulation of experimentally produced pain in man by electrical stimulation of some cortical, thalamic and basal ganglia structures. In: *Advances in Pain Research and Therapy*, edited by J. J. Bonica and D. Albe-Fessard. New York: Raven, 1976, vol. 1, p. 475–478.

55. Haber, L. H., R. F. Martin, J. M. Chung, and W. D. Willis. Inhibition and excitation of primate spinothalamic tract neurons by stimulation in region of nucleus reticularis gigantocellularis. *J. Neurophysiol.* 43: 1578–1593, 1980.
56. Hall, J. G., A. W. Duggan, S. M. Johnson, and C. R. Morton. Medullary raphe lesions do not reduce descending inhibition of dorsal horn neurones of the cat. *Neurosci. Lett.* 25: 25–29, 1981.
57. Hancock, M. B. Cells of origin of hypothalamo-spinal projections in the rat. *Neurosci. Lett.* 3: 179–184, 1976.
58. Handwerker, H. O., A. Iggo, and M. Zimmermann. Segmental and supraspinal actions on dorsal horn neurons responding to noxious and non-noxious skin stimuli. *Pain* 1: 147–165, 1975.
59. Hayes, R. L., P. G. Newlon, J. A. Rosecrans, and D. J. Mayer. Reduction of stimulation-analgesia by lysergic acid diethylamide, a depressor of serotonergic neural activity. *Brain Res.* 122: 367–372, 1977.
60. Hayes, R. L., D. D. Price, G. J. Bennett, G. L. Wilcox, and D. J. Mayer. Differential effects of spinal cord lesions on narcotic and non-narcotic suppression of nociceptive reflexes: further evidence for the physiologic multiplicity of pain modulation. *Brain Res.* 155: 91–101, 1978.
61. Hayes, R. L., D. D. Price, M. Ruda, and R. Dubner. Suppression of nociceptive responses in the primate by electrical stimulation of the brain or morphine administration: behavioral and electrophysiological comparisons. *Brain Res.* 167: 417–421, 1979.
62. Headley, P. M., A. W. Duggan, and B. T. Griersmith. Selective reduction by noradrenaline and 5-hydroxytryptamine of nociceptive responses of cat dorsal horn neurons. *Brain Res.* 145: 185–189, 1978.
63. Heath, R. G., and W. A. Mickle. Evaluation of seven years' experience with depth electrode studies in human patients. In: *Electrical Studies on the Unanesthetized Brain*, edited by E. R. Ramey and D. S. O'Doherty. New York: Hoeber, 1960, p. 214–247.
64. Hentall, I. D., and H. L. Fields. Segmental and descending influences on intraspinal thresholds of single C-fibers. *J. Neurophysiol.* 42: 1527–1537, 1979.
65. Hodge, C. J., Jr., A. V. Apkarian, R. Stevens, G. Vogelsang, and H. J. Wisnicki. Locus coeruleus modulation of dorsal horn unit responses to cutaneous stimulation. *Brain Res.* 204: 415–420, 1981.
66. Hökfelt, T., L. Terenius, H. G. J. M. Kuypers, and O. Dann. Evidence for enkephalin immunoreactive neurons in the medulla oblongata projecting to the spinal cord. *Neurosci. Lett.* 14: 55–60, 1979.
67. Hosobuchi, Y., J. E. Adams, and R. Linchitz. Pain relief by electrical stimulation of the central gray matter in humans and its reversal by naloxone. *Science* 197: 183–186, 1977.
68. Hosoya, Y. The distribution of spinal projection neurons in the hypothalamus of the rat, with the HRP method. *Exp. Brain Res.* 40: 79–87, 1980.
69. Jeftinija, S., K. Semba, and M. Randic. Norepinephrine reduces excitability of single cutaneous primary afferent C-fibers in the cat spinal cord. *Brain Res.* 219: 456–463, 1981.
70. Jordan, L. M., D. R. Kenshalo, R. F. Martin, L. H. Haber, and W. D. Willis. Depression of primate spinothalamic tract neurons by iontophoretic application of 5-hydroxytryptamine. *Pain* 5: 135–142, 1978.
71. Jurna, I. Effect of stimulation in the periaqueductal grey matter on activity in ascending axons of the rat spinal cord: selective inhibition of activity evoked by afferent A-δ and C fibre stimulation and failure of naloxone to reduce inhibition. *Brain Res.* 196: 33–42, 1980.

72. Krayniak, P. F., S. Weiner, and A. Siegel. An analysis of the efferent connections of the septal area in the cat. *Brain Res.* 189: 15–29, 1980.
73. Kuypers, H. G. J. M., and V. A. Maisky. Funicular trajectories of descending brain stem pathways in cat. *Brain Res.* 136: 159–165, 1977.
74. LeBars, D., A. H. Dickenson, and J. M. Besson. Diffuse noxious inhibitory controls (DNIC). II. Lack of effect on non-convergent neurones, supraspinal involvement and theoretical implications. *Pain* 6: 305–327, 1979.
75. Leichnetz, G. R., L. Watkins, G. Griffin, R. Murfin, and D. J. Mayer. The projection from nucleus raphe magnus and other brainstem nuclei to the spinal cord in the rat: a study using the HRP blue-reaction. *Neurosci. Lett.* 8: 119–124, 1978.
76. Lineberry, C. G., and C. J. Vierck. Attenuation of pain reactivity by caudate nucleus stimulation in monkeys. *Brain Res.* 98: 110–134, 1975.
77. Lovick, T. A., D. C. West, and J. H. Wolstencroft. Responses of raphespinal and other bulbar raphe neurones to stimulation of the periaqueductal gray in the cat. *Neurosci. Lett.* 8: 45–49, 1978.
78. Lovick, T. A., and J. H. Wolstencroft. Inhibitory effects of nucleus raphe magnus on neuronal responses in the spinal trigeminal nucleus to nociceptive compared with non-nociceptive inputs. *Pain* 7: 135–145, 1979.
79. Martin, R. F., L. H. Haber, and W. D. Willis. Primary afferent depolarization of identified cutaneous fibers following stimulation in medial brain stem. *J. Neurophysiol.* 42: 779–790, 1979.
80. Martin, R. F., L. M. Jordan, and W. D. Willis. Differential projections of cat medullary raphe neurons demonstrated by retrograde labelling following spinal cord lesions. *J. Comp. Neurol.* 182: 77–88, 1978.
81. Mayer, D. J., and J. C. Liebeskind. Pain reduction by focal electrical stimulation of the brain: an anatomical and behavioral analysis. *Brain Res.* 68: 73–93, 1974.
82. Mayer, D. J., and D. D. Price. Central nervous system mechanisms of analgesia. *Pain* 2: 379–404, 1976.
83. McCreery, D. B., and J. R. Bloedel. Reduction of the response of cat spinothalamic neurons to graded mechanical stimuli by electrical stimulation of the lower brainstem. *Brain Res.* 97: 151–156, 1975.
84. McCreery, D. B., J. R. Bloedel, and E. G. Hames. Effects of stimulating in raphe nuclei and in reticular formation on response of spinothalamic neurons to mechanical stimuli. *J. Neurophysiol.* 42: 166–182, 1979.
85. Necker, R., and R. F. Hellon. Noxious thermal input from the rat tail: modulation by descending influences. *Pain* 4: 231–242, 1978.
86. Nygren, L.-G., and L. Olson. A new major projection from locus coeruleus: the main source of noradrenergic nerve terminals in the ventral and dorsal columns of the spinal cord. *Brain Res.* 132: 85–93, 1977.
87. Oleson, T. D., D. B. Kirkpatrick, and S. J. Goodman. Elevation of pain threshold to tooth shock by brain stimulation in primates. *Brain Res.* 194: 79–95, 1980.
88. Oliveras, J. L., J. M. Besson, G. Guilbaud, and J. C. Liebeskind. Behavioral and electrophysiological evidence of pain inhibition from midbrain stimulation in the cat. *Exp. Brain Res.* 20: 32–44, 1974.
89. Oliveras, J. L., S. Bourgoin, F. Hery, J. M. Besson, and M. Hamon. The topographical distribution of serotoninergic terminals in the spinal cord of the cat: biochemical mapping by the combined use of microdissection and microassay procedures. *Brain Res.* 138: 393–406, 1977.
90. Oliveras, J. L., Y. Hosobuchi, F. Redjemi, G. Guilbaud, and J. M. Besson. Opiate antagonist, naloxone, strongly reduces analgesia induced by stimulation of a raphe nucleus (centralis inferior). *Brain Res.* 120: 221–229, 1977.
91. Oliveras, J. L., F. Redjemi, G. Guilbaud, and J. M. Besson. Analgesia induced by

electrical stimulation of the inferior centralis nucleus of the raphe in the cat. *Pain* 1: 139–145, 1975.

92. Petrovicky, P., O. Kadlecova, and K. Masek. Mutual connections of the raphe system and hypothalamus in relation to fever. *Brain Res. Bull.* 7: 131–149, 1981.
93. Price, D. D., and R. Dubner. Neurons that subserve the sensory-discriminative aspects of pain. *Pain* 3: 307–338, 1977.
94. Proudfit, H. K. Reversible inactivation of raphe magnus neurons: effects on nociceptive threshold and morphine-induced analgesia. *Brain Res.* 201: 459–464, 1980.
95. Proudfit, H. K., A. A. Larson, and E. G. Anderson. The role of GABA and serotonin in the mediation of raphe-evoked spinal cord dorsal root potentials. *Brain Res.* 195: 149–165, 1980.
96. Randic, M., and H. H. Yu. Effects of 5-hydroxytryptamine and bradykinin in cat dorsal horn neurones activated by noxious stimuli. *Brain Res.* 111: 197–203, 1976.
97. Reynolds, D. V. Surgery in the rat during electrical analgesia induced by focal brain stimulation. *Science* 164: 444–445, 1969.
98. Rhodes, D. L. Periventricular system lesions and stimulation-produced analgesia. *Pain* 7: 51–63, 1979.
99. Rhodes, D. L., and J. C. Liebeskind. Analgesia from rostral brain stem stimulation in the rat. *Brain Res.* 143: 521–532, 1978.
100. Richardson, D. E., and H. Akil. Pain reduction by electrical brain stimulation in man. Pt. 1. Acute administration in periaqueductal and periventricular sites. *J. Neurosurg.* 47: 178–183, 1977.
101. Rivot, J. P., A. Chaouch, and J. M. Besson. The influence of naloxone on the C fiber response of dorsal horn neurons and their inhibitory control by raphe magnus stimulation. *Brain Res.* 176: 355–364, 1979.
102. Rivot, J. P., A. Chaouch, and J. M. Besson. Nucleus raphe magnus modulation of response of rat dorsal horn neurons to unmyelinated fiber inputs: partial involvement of serotonergic pathways. *J. Neurophysiol.* 44: 1039–1057, 1980.
103. Sakuma, Y., and D. W. Pfaff. Convergent effects of lordosis-relevant somatosensory and hypothalamic influences on central gray cells in the rat mesencephalon. *Exp. Neurol.* 70: 269–281, 1980.
104. Sandberg, D. E., and M. Segal. Pharmacological analysis of analgesia and self-stimulation elicited by electrical stimulation of catecholamine nuclei in the rat brain. *Brain Res.* 152: 529–542, 1978.
105. Satoh, M., A. Akaike, T. Nakazawa, and H. Takagi. Evidence for involvement of separate mechanisms in the production of analgesia by electrical stimulation of the nucleus reticularis paragiganto-cellularis and nucleus raphe magnus in the rat. *Brain Res.* 194: 525–529, 1980.
106. Schmidek, H. H., D. Fohanno, F. R. Ervin, and W. H. Sweet. Pain threshold alterations by brain stimulation in the monkey. *J. Neurosurg.* 35: 715–722, 1971.
107. Sessle, B. J., J. W. Hu, R. Dubner, and G. E. Lucier. Functional properties of neurons in cat trigeminal subnucleus caudalis (medullary dorsal horn). II. Modulation of responses to noxious and nonnoxious stimuli by periaqueductal gray, nucleus raphe magnus, cerebral cortex, and afferent influences, and effect of naloxone. *J. Neurophysiol.* 45: 193–207, 1981.
108. Steinbusch, H. W. M. Distribution of serotonin-immunoreactivity in the central nervous system of the rat-cell bodies and terminals. *Neuroscience* 6: 557–618, 1981.
109. Swanson, L. W., and W. M. Cowan. The connections of the septal region in the rat. *J. Comp. Neurol.* 186: 621–656, 1979.
110. Taub, A. Local, segmental and supraspinal interaction with a dorsolateral spinal cutaneous afferent system. *Exp. Neurol.* 10: 357–374, 1964.

111. Tohyama, M., K. Sakai, D. Salvert, M. Touret, and M. Jouvet. Spinal projections from the lower brain stem in the cat as demonstrated by the horseradish peroxidase technique. I. Origins of the reticulospinal tracts and their funicular trajectories. *Brain Res.* 173: 383–403, 1979.
112. Wall, P. D. The laminar organization of dorsal horn and effects of descending impulses. *J. Physiol. London* 188: 403–423, 1967.
113. Willis, W. D., L. H. Haber, and R. F. Martin. Inhibition of spinothalamic tract cells and interneurons by brain stem stimulation in the monkey. *J. Neurophysiol.* 40: 968–981, 1977.
114. Yaksh, T. L., and T. A. Rudy. Narcotic analgetics: CNS sites and mechanisms of action as revealed by intracerebral injection techniques. *Pain* 4: 299–359, 1978.
115. Yezierski, R. P., T. K. Wilcox, and W. D. Willis. A pharmacological study of inhibition of primate spinothalamic tract cells from periaqueductal gray and nucleus raphe magnus. *Soc. Neurosci. Abstr.* 6: 40, 1980.
116. Yokota, T., and S. Hashimoto. Periaqueductal gray and tooth pulp interaction on units in caudal medulla oblongata. *Brain Res.* 117: 508–512, 1976.
117. Zimmermann, M. Encoding in dorsal horn interneurons receiving noxious and non-noxious afferents. *J. Physiol. Paris* 73: 221–232, 1978.
118. Zimmermann, M. Somatovisceral sensations: processing in the central nervous system. In: *Human Physiology*, edited by R. F. Schmidt and G. Thews. Berlin: Springer-Verlag, in press.
119. Zorman, G., I. D. Hentall, J. E. Adams, and H. L. Fields. Naloxone reversible analgesia produced by microstimulation in the rat medulla. *Brain Res.* 219: 137–148, 1981.

SIX

Stimulation-Produced Analgesia

Thomas L. Wolfle
Interagency Primate Steering Committee, National Institutes of Health, Besthesda, Maryland

John C. Liebeskind
Department of Psychology, University of California, Los Angeles, California

Although it is unlikely that stimulation-produced analgesia (SPA) will soon become a clinical tool in veterinary medicine, animals continue to be valuable in studies that provide information about an endogenous pain-control system of ultimate benefit to humans and to animals. Many milestones in science are accomplished paradoxically as with chronic pain where many advances have come from behavioral models of acute pain. Fortuitously, SPA differs somewhat between humans and animals. Although in humans medial brain stem stimulation alleviates pathological pain without affecting normal (acute) pain perception, the latter is typically used to define SPA in animals. Also the duration of SPA effectiveness in humans greatly exceeds that in animals. Whether these are true species differences or result from differences in techniques of stimulation or pain measurement is not yet known. Although animals can not verbally communicate their pain, their behavior is similar to that of humans in response to noxious stimulation. The most basic of these behaviors is the escape response, which removes the body or limb from threat of tissue destruction. A common demonstration of SPA is that a mouse on a hot plate or a rat with its tail under a heat lamp fails to make the basic escape response.

Chronic pain is an enigma because, unlike acute pain which signals tissue damage, it often appears to serve no useful biological purpose. It is a sensation of paramount importance for which the search for relief often supplants all other drives. Social interactions may cease and many biological functions are altered with chronic pain. Personalities change, family ties may be undermined, and jobs are threatened—all for the lack of relief. Often continuing long after an injury has healed, the afflicted victim may go to great ends including repeated surgery to find relief. These searches may lead to lives numbed by alcohol and drugs and even to the ultimate escape—suicide. Bonica (6) estimates that 85 million patients suffer from unrelenting pain associated with chronic diseases including cancer, headaches, back and musculoskeletal disorders, and arthritis. In the United States alone, millions of work days are lost, which together with medical expenses total some 60 billion dollars annually.

The differences between acute and chronic pain are profound. Although treatment with analgesics, muscle relaxants, and tranquilizers generally alleviates acute pain, such treatment is usually ineffective for chronic pain. Even neurosurgery, which has been the final hope for chronic pain sufferers, may carry the risk of producing new pain and is often reserved only for terminally ill patients. The ineffectiveness of drugs and surgery to combat chronic pain led researchers away from these traditional approaches to a new direction. Spurred on by the hypothesis of Melzack and Wall (35) that the body possesses an autonomous descending control from higher centers to spinal gates, investigators looked for ways to activate these gates and block pain impulses from reaching the higher centers. The notion that electrical impulses might serve in this manner draws support from a wide variety of scientific and anecdotal evidence dating back to Aristotle's *Historia Animalium*, in which he described a numbing sensation resulting from touching an electric ray (*Torpedo*).

Given the magnitude of the problem of chronic pain and the disappointment in traditional treatment, it is not surprising that much recent attention has been focused on understanding the body's own defense against pain. Of the articles printed in scientific journals in 1976, 20 of the 100 most frequently cited articles in 1977 dealt with endogenous opioid peptides (12), the newly discovered neuropeptides quickly labeled "the brain's own morphine." The importance of this is clearer when one realizes that the first of these peptides (the enkephalins) was described only 1 year earlier (21, 22), and it was just over a decade since it was first hypothesized from studies in the rat that the brain possesses an endogenous pain-inhibitory system (33).

There is now much evidence for a central nervous system (CNS) mechanism that functions to block pain through the activation of descending controls. The brain substrate responsible for this pain-inhibitory system, which appears to use opioid peptides as chemical mediators, was mapped through studies in which electrical stimulation of specific brain loci produced behavioral evidence of analgesia. It was learned that direct electrical stimulation of portions of the medial brain stem of the awake rat (31, 33), monkey (14), and cat (38) caused powerful inhibition of escape responses to painful stimuli as well as to the perception of pain itself.

The specificity of this inhibition is indeed quite remarkable. Electrical stimulation of tooth pulp is a common laboratory procedure for producing an operationally well-defined painful stimulus. In addition to other musculoskeletal and autonomic manifestations of pain, this stimulus elicits a reflexive jaw-opening response. Stimulation-produced analgesia has been shown to block this jaw-opening response, although the same response elicited by an innocuous tap on the tooth is unaffected (39). Also, Mayer et al. (33) demonstrated in the rat that SPA occurs without deficits in other modalities. These studies have been important in elucidating the mechanism inhibiting pain responses. It is now clearly established that SPA affects only pain and not the animal's ability to respond. The term *analgesia* is used in this chapter to conform with the literature; however, the term *hypalgesia* may be more accurate. Specific electrode placements of SPA define a discrete area of peripheral analgesia, outside of which normal pain responses can be elicited (31, 33).

In addition to blocking pain responses in animals to electric shock (33), heating of the skin (31), subcutaneous injections of formaldehyde (34), and intraperitoneal injections of hypertonic saline (13), SPA has been described electrophysiologically in brain and spinal neuron recordings (36–38, 44, 45). Multiple-unit activity (MUA), recorded from the periaqueductal gray (PAG) matter and other SPA sites, demonstrates remarkably similar increases in activity from either SPA or analgesic doses of morphine (37). In areas not supporting SPA, MUA is either diminished or unaffected by morphine (45, but cf. 11).

In 1969, Reynolds (40) first convincingly demonstrated the power of SPA. He reported that stimulation of the PAG of rats produced analgesia profound enough to permit abdominal surgery without anesthesia. Two years later, we suggested that SPA and opiate analgesia shared both sites and mechanisms of action (33), and more recent evidence adds credence to this hypothesis. Electrical stimulation and microinjections of morphine into the PAG cause analgesia (48). The similarity between mechanisms of SPA and opiate analgesia has frequently been cited, and the potency of SPA is reported comparable to that achieved by high doses of morphine (31, 33). Likewise, lesions of the dorsolateral funiculus of the spinal cord block both SPA and morphine analgesia (5). This finding emphasizes the critical role of this descending path in providing modulation from the medial brain stem to the spinal cord dorsal horn.

Patients receiving repeated doses of morphine soon become tolerant and require higher doses to achieve the same degree of pain relief.

Mayer and Hayes (30) argued that if morphine and SPA act through the same pain-blocking mechanism, tolerance induced by either morphine or SPA should render the other one tolerant as well (i.e., cross-tolerant). Indeed they demonstrated that not only tolerance develops from repeated SPA trials, but cross-tolerance between SPA and morphine also occurs. Hosobuchi et al. (18) found that candidate patients for SPA have frequently developed tolerance to morphine to the extent that SPA produces little pain relief. To prevent this cross-tolerance, the essential amino acid L-tryptophan is now administered to the patients for several weeks before surgery until reversal to morphine tolerance is seen. Low-level maintenance doses of this drug are normally found effective in preventing tolerance to SPA, possibly by increasing β-endorphin levels (18), which are often low in chronic pain patients (4).

Shortly after Mayer et al. (33) hypothesized the existence of an endogenous or intrinsic pain-inhibitory system triggered by administration of opiate drugs or PAG stimulation, Akil et al. (2, 3) demonstrated in rats that the opiate antagonist naloxone could at least partly block SPA. This observation supported the view that SPA and opiate drugs share a common mechanism of action. Several recent studies have failed to confirm naloxone blockade (cf. 49). However, it now appears that two different SPA sites coexist in the PAG: a naloxone-sensitive ventral region and a more dorsal naloxone-insensitive region (25). Although binding sites other than those for opiates are clearly involved, the finding of opiate binding sites and opiate peptide-containing cell bodies in brain areas supporting SPA areas also sensitive to microinjection of opioids (25, 29, 48) leaves little doubt that endogenous opiates play a role in pain suppression.

Recently, efforts have been directed at discovering under what natural conditions the endogenous analgesia system is triggered. Liebeskind et al. (27) suggested from studies in primates that this substrate would not be accessed trivially because aversive stimuli provide the opportunity for adaptive responses that have important survival value. However, intensely noxious or stressful situations can interfere with effective coping responses, and in this case activation of the intrinsic analgesia system would prove adaptive. Thus Akil et al. (1) and Hayes et al. (15) demonstrated that stress could have a powerful analgesic effect in the rat, and hence stress is a likely environmental trigger for endogenous opiate release. However, just as naloxone has been found to block some but not all SPA, suggesting both opiate- and nonopiate-mediated analgesic systems, these workers reported different results with naloxone administration during stress analgesia. The stress analgesia studied by Akil et al. was blocked by naloxone, but that studied by Hayes et al. was not. It was recently shown in rats that depending on the temporal parameters of inescapable footshock or area of the body stimulated (47a), stress analgesia is or is not blocked by naloxone (23, 47a). Other paradigms have shown similar differences. Rats previously exposed to escapable or inescapable shock differed from one another 24 h later when tested with either the hot plate or tail-flick method (28). Maier and Jackson (28) reported that brief reexposure to shock just before testing induced analgesia only in rats previously subjected to inescapable shock. Also the preference that rats normally show for signaled versus unsignaled footshock was reported to be blocked by naloxone (10). Such studies can only hint at directions for future work, however, because of the difficulty in relating studies, in which subtle differences in test parameters, diet, handling, individual differences in pain sensitivity (7), or previous experience undoubtedly play a role.

The system most frequently associated with stress is the pituitary-adrenal axis, which is generally thought to play an essential role in Selye's (43) stress adaptation response and may be involved in stress analgesia. It has been shown that stress causes the release of β-endorphin from the pituitary (42); hypophysectomy reduces opioid stress analgesia (24); and adrenalectomy, adrenal demedullation, and adrenal medulla denervation (celiac ganglionectomy) all block opioid stress analgesia (26). Nonopioid stress analgesia was unaffected in these studies. Lewis et al. (25) concluded from these adrenal lesion studies that the blocking of stress analgesia was attributable to a reduction in enkephalinlike peptides from the adrenal medulla. To support this view, they provide evidence that reserpine, known to augment both the adrenal content and stimulation-induced release of enkephalins (46), increases opioid stress analgesia significantly (26).

Because of the inherent problems associated with long-term administration of opiate drugs, there is considerable value in further understanding the nonopiate system as well as the nonopiate synaptic transmitters within the opiate system. In this regard Lewis et al. (25) recently demonstrated that the cholinergic muscarinic antagonist scopolamine, but not the centrally inactive methylscopolamine, blocks the opioid but not the nonopioid form of stress analgesia. This suggests that acetylcholine stimulates release of opioid peptides and that cholinergic synapses exist in the central opioid pathway. Further evidence of this comes from studies in mice demonstrating that oxotremorine, a cholinergic agonist, causes analgesia sensitive to blockade by naloxone (20).

Although not directly relevent to a review of SPA, an example of promising clinically relevant technology resulting in part from the functional descriptions of endogenous opiates provided by animal models of SPA is the discovery of naloxone as an agent in pain relief. Realizing that some severe stress produces a state of shock characterized by hypotension, which is mediated by endorphin and blocked by naloxone (16), investigators have recently suggested that naloxone administration may facilitate recovery from spinal cord injuries that often leave victims paralyzed and in pain for life. Hypothesizing that endorphins are released after spinal cord injury, Faden et al. (9) suggested that they exacerbate resulting spinal cord ischemia by causing reduction in spinal cord blood flow. Thus naloxone should increase blood flow by blocking the opiate effect and hence reduce secondary injury caused by ischemia. Recent evidence tends to support this hypothesis even when treatment is delayed 4 h after injury (8, 9, 50).

Pain is difficult to study, because in the laboratory it must be created under strict ethical guidelines (7a). However, humans have begun to benefit from these SPA studies in animals, finding relief for the first time in years (17, 19, 41).

REFERENCES

1. Akil, H., J. Madden, R. L. Patrick, and J. D. Barchas. Stress-induced increase in endogenous opiate peptides: concurrent analgesia and its partial reversal by naloxone. In: *Opiates and Endogenous Opioid Peptides*, edited by H. W. Kosterlitz. Amsterdam: Elsevier, 1976, p. 63–70.
2. Akil, H., D. J. Mayer, and J. C. Liebeskind. Comparison chez le rat entre l'analgesia induite par stimulation de la substance grise periaqueducal el l'analgesie morphinique. *C. R. Acad. Sci.* 274: 3603–3605, 1972.
3. Akil, H., D. J. Mayer, and J. C. Liebeskind. Antagonism of stimulation-produced analgesia by naloxone, a narcotic antagonist. *Science* 191: 961–962, 1976.
4. Akil, H., S. J. Watson, P. A. Berger, and J. D. Barchas. Endorphins, β-LPH, and ACTH: biochemical, pharmacological, and anatomical studies. In: *The Endorphins*, edited by E. Costa and M. Trabucchi. New York: Raven, 1978, p. 125–140.
5. Basbaum, A. I., N. J. Marley, J. O'Keefe, and C. H. Clanton. Reversal of morphine and stimulus-produced analgesia by subtotal spinal cord lesions. *Pain* 3: 43–56, 1977.
6. Bonica, J. J. Pain research and therapy: past and current status and future trends. In: *Pain Discomfort and Humanitarian Care*, edited by L. K. Y. Ng and J. J. Bonica. New York: Elsevier/North-Holland, 1979, p. 1–46.
7. Chance, W. T., A. C. White, G. M. Krynock, and J. A. Rosecrans. Conditional fear-induced antinociception and decreased binding of [^{3}H]N-leu-enkephalin to rat brain. *Brain Res.* 141: 371–374, 1978.

7a. Ethical standards for investigations of experimental pain in animals. The Committee for Research and Ethical Issues of the International Association for the Study of Pain (Editorial). *Pain* 9: 141–143, 1980.

8. Faden, A. I., T. P. Jacobs, and J. W. Holaday. Opiate antagonist improves neurologic recovery after spinal injury. *Science* 211: 493–494, 1981.
9. Faden, A. I., T. P. Jacobs, E. Mougrey, and J. W. Holaday. Endorphins in experimental spinal injury: therapeutic effect of naloxone. *Neurology* 10: 326–332, 1981.

10. Fanselow, M. S. Naloxone attenuates rat's preference for signaled shock. *Physiol. Psychol.* 7: 70–74, 1979.
11. Frederickson, R. C., and F. H. Norris. Enkephalin-induced depression of single neurons in brain areas with opiate receptors—antagonism by naloxone. *Science* 194: 440–442, 1976.
12. Garfield, E. Current comments: the 1976 articles most cited in 1976 and 1977. Pt. I. Life sciences. In: *Essays of an Information Scientist*, edited by E. Garfield. Philadelphia, PA: ISI, 1979–1980, vol. 4, p. 89–91.
13. Giesler, G. J., Jr., and J. C. Liebeskind. Inhibition of visceral pain by electrical stimulation of the periaqueductal gray matter. *Pain* 2: 43–48, 1976.
14. Goodman, S. J., and V. Holcombe. Selective and prolonged analgesia in monkey resulting from brain stimulation. In: *Advances in Pain Research and Therapy*, edited by J. J. Bonica and D. Albe-Fessard. New York: Raven, 1976, vol. 1, p. 495–502.
15. Hayes, R. L., G. J. Bennett, P. G. Newlon, and D. J. Mayer. Behavioral and physiological studies of non-narcotic analgesia in the rat elicited by certain environmental stimuli. *Brain Res.* 155: 69–90, 1978.
16. Holaday, J. W., and A. I. Faden. Naloxone reversal of endotoxin hypotension suggests role of endorphins in shock. *Nature London* 275: 450–451, 1978.
17. Hosobuchi, Y., J. E. Adams, and R. Linchitz. Pain relief by electrical stimulation of the central gray matter in humans and its reversal by naloxone. *Science* 197: 183–186, 1977.
18. Hosobuchi, Y., J. Rossier, and F. E. Bloom. Oral loading with L-tryptophan may augment the simultaneous release of ACTH and beta-endorphin that accompanies periaqueductal stimulation in humans. In: *Neural Peptides and Neural Communications*, edited by E. Costa and M. Trabucchi. New York: Raven, 1980, p. 563–569.
19. Hosobuchi, Y., J. Rossier, F. E. Bloom, and R. Guillemin. Stimulation of human periaqueductal gray matter for pain relief increases immunoreactive β-endorphin in ventricular fluid. *Science* 203: 279–281, 1979.
20. Howes, J. F., L. S. Harris, W. L. Dewey, and C. A. Voyda. Brain acetylcholine levels and inhibition of the tail-flick reflex in mice. *J. Pharmacol. Exp. Ther.* 169: 23–28, 1969.
21. Hughes, J. Isolation of an endogenous compound from the brain with pharmacological properties similar to morphine. *Brain Res.* 88: 295–308, 1975.
22. Hughes, J., T. W. Smith, H. W. Kosterlitz, L. A. Fothergill, B. A. Morgan, and H. R. Morris. Identification of two related pentapeptides from the brain with potent opiate agonist activity. *Nature London* 258: 577, 1975.
23. Lewis, J. W., J. T. Cannon, and J. C. Liebeskind. Opioid and nonopioid mechanisms of stress analgesia. *Science* 208: 623–625, 1980.
24. Lewis, J. W., E. H. Chudler, J. T. Cannon, and J. C. Liebeskind. Hypophysectomy differentially affects morphine and stress analgesia. *Proc. West. Pharmacol. Soc.* 24: 323–326, 1981.
25. Lewis, J. W., M. S. Stapleton, A. J. Castiglioni, and J. C. Liebeskind. Stimulation-produced analgesia and intrinsic mechanisms of pain suppression. In: *Neuropeptides—Basic and Clinical Aspects*, edited by G. Fink and L. J. Whalley. Edinburgh: Churchill Livingstone, 1982, p. 41–49. (Proc. 11th Pfizer Int. Symp.)
26. Lewis, J. W., M. G. Tordoff, J. E. Sherman, and J. C. Liebeskind. Adrenal medullary enkephalin-like peptides may mediate opioid stress analgesia. *Science* 217: 557–559, 1982.
27. Liebeskind, J. C., G. J. Giesler, and G. Urca. Evidence pertaining to an endogenous mechanism of pain inhibition in the central nervous system. In: *Sensory Functions of the Skin in Primates*, edited by Y. Zotterman. Oxford, UK: Pergamon, 1976, p. 561–573.

28. Maier, S. F., and R. L. Jackson. Learned helplessness: all of us were right (and wrong), inescapable shock has multiple effects. In: *Psychology of Learning and Motivation*, edited by G. H. Bower. New York: Academic, 1979, vol. 13, p. 155–214.
29. Mayer, D. J. Endogenous analgesia systems:neural and behavioral mechanisms. In: *Advances in Pain Research and Therapy*, edited by J. J. Bonica, J. C. Liebeskind, and D. G. Albe-Fessard. New York: Raven, 1979, vol. 3, p. 385–410.
30. Mayer, D. J., and R. L. Hayes. Stimulation-produced analgesia: development of tolerance and cross-tolerance to morphine. *Science* 188: 941–943, 1975.
31. Mayer, D. J., and J. C. Liebeskind. Pain reduction by focal electrical stimulation of the brain: an anatomical and behavioral analysis. *Brain Res.* 68: 73–93, 1974.
33. Mayer, D. J., T. L. Wolfle, H. Akil, B. Carder, and J. C. Liebeskind. Analgesia from electrical stimulation in the brainstem of the rat. *Science* 174: 1351–1354, 1971.
34. Melzack, R., and D. F. Melinkoff. Analgesia produced by brain stimulation: evidence of a prolonged onset period. *Exp. Neurol.* 43: 369–574, 1974.
35. Melzack, R., and P. D. Wall. Pain mechanisms: a new theory. *Science* 150: 971–979, 1965.
36. Morrow, T. J., and K. L. Casey. Analgesia produced by mesencephalic stimulation: effect on bulboreticular neurons. In: *Advances in Pain Research and Therapy*, edited by J. J. Bonica and D. Albe-Fessard. New York: Raven, 1976, vol. 1, p. 503–510.
37. Oleson, T. D., D. A. Twombly, and J. C. Liebeskind. Effects of pain attenuating brain stimulation and morphine on electrical activities in the raphe nuclei of the awake rat. *Pain* 4: 211–230, 1978.
38. Oliveras, J. L., J. M. Besson, G. Guilbaud, and J. C. Liebeskind. Behavioral and electrophysiological evidence of pain inhibition from midbrain stimulation in the cat. *Exp. Brain Res.* 20: 32–44, 1974.
39. Oliveras, J. L., A. Woda, G. Guilbaud, and J. M. Besson. Inhibition of the jaw opening reflex by electrical stimulation of the periaqueductual gray matter in the awake, unrestrained cat. *Brain Res.* 72: 328–331, 1974.
40. Reynolds, D. V. Surgery in the rat during electrical analgesia induced by focal brain stimulation. *Science* 164: 444–445, 1969.
41. Richardson, D. E., and H. Akil. Pain reduction by electrical brain stimulation in man. *J. Neurosurg.* 47: 178–189, 1977.
42. Rossier, J., E. D. French, C. Rivier, N. Ling, R. Guillemin, and F. E. Bloom. Foot-shock induced stress increases β-endorphin levels in blood but not brain. *Nature London* 270: 618–620, 1977.
43. Selye, H. *The Stress of Life.* New York: McGraw-Hill, 1976.
44. Urca, G., R. L. Nahin, and J. C. Liebeskind. Development of tolerance to the effects of morphine: association between analgesia and electrical activity in the periaqueductal gray matter. *Brain Res.* 176: 202–207, 1979.
45. Urca, G., R. L. Nahin, and J. C. Liebeskind. Effects of morphine on spontaneous multiple unit activity: possible relation to mechanisms of analgesia and reward. *Exp. Neurol.* 66: 248–262, 1979.
46. Viveros, O. H., E. J. Diliberto, Jr., E. Hazum, and K.-J. Chang. Enkephalins as possible adrenomedullary hormones: storage, secretion and regulation of synthesis. In: *Neural Peptides and Neuronal Communications*, edited by E. Costa and M. Trabucchi. New York: Raven, 1980, p. 191–204.
47. Wall, P. D., and M. Gutnick. Ongoing activity in peripheral nerves: the physiology and pharmacology of impulses originating from a neuroma. *Exp. Neurol.* 43: 580–593, 1974.
47a. Watkins, L. R., D. J. Mayer. Organization of endogenous opiate and nonopiate pain control systems. *Science* 216: 1185–1192, 1982.

48. Yaksh, T. L., and T. A. Rudy. Narcotic analgesia: CNS sites and mechanisms of action as revealed by intracerebral injection techniques. *Pain* 4: 299–359, 1978.
49. Yaksh, T. L., J. C. Yeung, and T. A. Rudy. An inability to antagonize with naloxone the elevated thresholds resulting from electrical stimulation of the mesencephalic central gray. *Life Sci.* 18: 1193–1198, 1976.
50. Young, W., E. S. Flamm, H. B. Demopulos, J. J. Tomasula, and V. DeCrescito. Effect of naloxone on post-traumatic ischemia in experimental spinal contusion. *J. Neurosurg.* 55: 209–219, 1981.

SEVEN

Human and Nonhuman Primate Reactions to Painful Electrocutaneous Stimuli and to Morphine

Charles J. Vierck, Jr.
Brian Y. Cooper
Richard H. Cohen
Department of Neuroscience and Center for Neurobiological Sciences, University of Florida College of Medicine, Gainesville, Florida

Indices of Pain Sensations and Reactions • Escape Latency as a Measure of Pain and Morphine Hypalgesia • Control for Response Suppression by Morphine • Importance of the Source of Experimental Pain • Summary

The difficulties inherent in the process of discriminating the presence of pain and its intensity in another individual are similar when evaluating nonhumans or humans. Humans have the advantage of utilizing unique verbal descriptors of pain intensity and quality, but reliance solely on verbal descriptors can be misleading. For example, an individual can verbally deny pain and yet display somatic and autonomic motor signs of nociception. We give special credence to verbal reports of pain sensations because the precision and efficiency of language is appropriate and necessary for qualitative distinctions. Verbal reports are advantageous for defining the spatial location and extent of a sensation and are necessary for distinguishing between varieties of pain such as sharp, dull, burning, pricking, or itching. However, the magnitudes of somatic and autonomic motor activities can reflect pain intensity, and quantification of these responses can be accomplished in a variety of animals. The challenge of comparative algesimetry is to establish measures of pain reactivity that are valid and reliable indicators of pain intensity.

Indices of Pain Sensations and Reactions

Humans can identify four component sensations that are elicited by brief (50-ms) trains of focal, electrical stimulation of the lateral calf (14). *1*) A nonpainful tap can be highly localized in space and time and is identifiable as resulting from an oscillating electrical source. This is the lowest-threshold sensation (T = 0.01 mA/mm^2) when current is varied systematically. *2*) A stinging, painful sensation is noticed at ~0.5 mA/mm^2, and the intensity of this brief event increases regularly with stimulus intensity. The stinging sensation is highly focal and commands attention quite effectively. The onset is immediate and abrupt, and the sensation disappears within 1 s of stimulus offset. *3*) A thermal component can appear at intensities above threshold for stinging pain elicited by alternating current. Heat sensations increase at a low slope with current and do not contribute substantially to pain intensity at the highest values that can be tolerated by normal subjects. The thermal sensation apparently depends on amount of current per unit of time and may result from heating the tissue at the electrode. *4*) At a relatively high threshold, similar to that of the thermal cue (at ~0.9 mA/mm^2), a delayed, second pain sensation appears and grows steadily with current magnitude. At 2.1 mA/mm^2, humans report a strong but tolerable first pain sensation that lasts ~0.5 s and is succeeded by a weaker and more gradually incrementing and decrementing second pain sensation.

Having identified an appropriate range of stimuli producing weak to strong sensations of rapid-onset, tolerable pain, it is feasible to describe functions relating sensation magnitude to stimulus intensities within a substantial but acceptable portion of the pain sensitivity range. Human subjects rated pain intensity by magnitude estimation, drawing lines with lengths proportional to pain intensity; the exponent of the power function fitting the data was 0.96. When the same subjects pulled a manipulandum to terminate (escape) the electrocutaneous stimuli, operant force was related to stimulus intensity with an exponent of 0.75. Thus the vigor of an adaptive motor response can reflect the intensity of painful stimuli, but different motor responses, different temporal relationships between stimulus and response, and other factors influence the growth function. It is not necessary or even expected that immediate corrective reactions to pain would be related to stimulus intensity with an exponent identical to that relating an introspective, posthoc evaluation of pain intensity to stimulus intensity.

Monkeys also can be trained to accept and readily tolerate electrical stimuli within the range of 0.6 to 2.1 mA/mm^2 (for details of the procedure see ref. 13). When the monkeys pulled the same manipulandum to escape shock (presented in 50-ms trains at 4 trains/s until a response occurred or until 5 s elapsed), the averaged exponents of the power functions relating operant force to stimulus intensity ranged from 0.35 to 0.54 in six experiments involving three species of monkey (Fig. 1). Also, different individual animals within a species approached the escape task with different strategies and produced individualistic stimulus-response functions. Therefore the slope of an averaged power function should not be regarded literally (7). Fortunately the purpose of relating reaction measures to stimulus intensity is not to show that all species regulate output in proportion to input according to the same formula for any or all response measures. Rather, the form of a stimulus-response relationship (e.g., a power function) and its slope can tell us *1*) that the measure is highly sensitive (high slope) or insensitive (flat) and *2*) whether we have exceeded the range of stimuli that produces a linear relationship. The monkeys' response functions

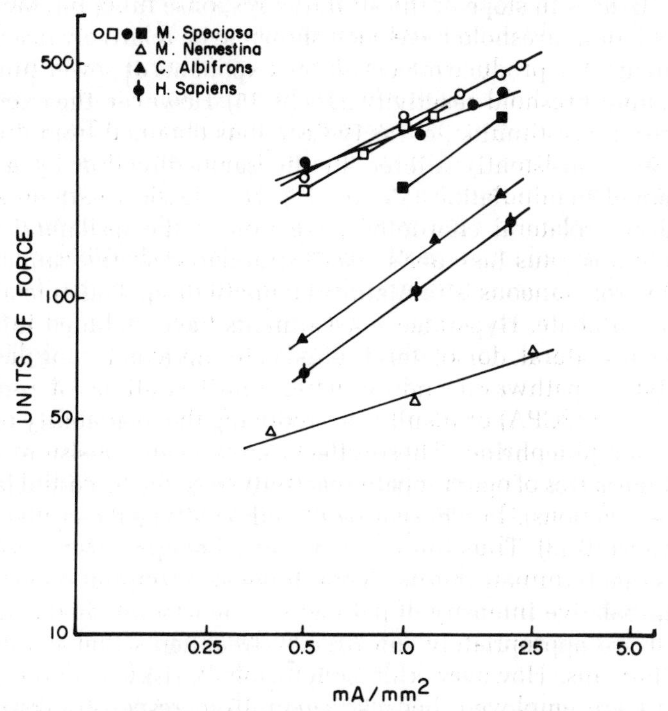

Fig. 1. Readouts of operant escape force are presented from 7 experiments involving stimulation of either lateral calf with 50-ms trains of alternating current or 200-Hz square pulses 0.5 ms in duration. Force of lever pulling or panel pushing was derived from voltage output of Statham gold cell transducer in series with 100-lb load cell. Output of force transducer was calibrated with weights, and function was linear within range of forces produced by animals. To separate functions on graph, units on ordinate represent transducer output in some cases and units of force in others; therefore slopes, but not absolute values, are comparable. In each experiment, slopes were $\gg 0$ but <1.

did not plateau at the high intensities, supporting the observation that the entire pain sensitivity range is not covered by the intensities utilized. That is, maximal forces and minimal latencies are likely to be produced by the monkeys at or below the upper limit of tolerable pain intensities. This is an important consideration in designing a humane paradigm of pain testing.

Realistic and crucial goals of a laboratory animal model of pain sensitivity are that each individual's functions relating response magnitude to stimulus intensity are stable with repeated testing under normal conditions and yet are sensitive to putative hypalgesic and hyperalgesic treatments (13). These goals are best achieved by observing reactions to weak and strong pain, permitting detection of increases or decreases in slope or the overall magnitude of functions related to suprathreshold stimuli. In contrast, threshold measures can be insensitive to changes in slope of the stimulus-response function. Measures of pain-reaction threshold have been shown to be relatively insensitive to treatments that produce effects on the exponents of power functions fitted to suprathreshold-reactivity data (9, 15). However, the exponents of pain-reaction, stimulus-intensity functions obtained from different animals are consistently altered in the same direction by a given experimental manipulation (13, 16, 17). Hypalgesic treatments have included anterolateral chordotomy, section of the ipsilateral dorsal spinal columns (plus Lissauer's tract), stimulation of the caudate nucleus, electrocutaneous stimulation (acupuncture), and injection of morphine sulphate. Hyperalgesic treatments have included interruption of the ipsilateral, dorsolateral spinal columns (containing descending inhibitory pathways) and systemic administrations of *p*-chlorophenylalanine (PCPA) or disulfiram, reducing the availability of serotonin or norepinephrine. These effects have been consistent across different measures of operant pain reactivity (e.g., the force and latency of escape reactions), but measures of reflexive vigor can give quite different results (3). Thus the vigor and speed of operant nonreflexive reactions that terminate strong electrocutaneous stimulation appear to reveal the relative intensity of pain according to individualized functions that are appropriately sensitive to treatments that manipulate pain in humans. However, this conclusion is risky when systemic treatments are employed, because generalized response suppression can masquerade as a specific effect on pain.

Escape Latency as a Measure of Pain and Morphine Hypalgesia

A basic assumption of most algesimetric methods is that each subject will choose to terminate painful stimuli and will tolerate long durations of stimuli below pain threshold. Estimates of this response tendency are based on some measure of response latency: *1*) a binary classification of latencies expressed as the percentage of trials on which a response occurs within an arbitrary trial duration, *2*) the average latency of escape responses, or *3*) a combined average of escape latencies and the durations of stimuli not escaped. Because subjects differentiate stimulus intensities by the percentage and latency of escape responses, the combined measure (trial duration) is desirable, but the arbitrary specification of maximal trial length constitutes a powerfully distorting influence (8). This is seen in Figure 2, which shows histograms of the percentage of escape responses emitted in 100-ms bins of time after onset of electrical stimulation to the lateral calf of macaque monkeys. These animals emitted at least 90% of their escape responses within 800 ms of trial onset, for shock intensities ranging from near pain threshold for humans (and escape threshold for the monkeys; 0.6 mA/mm^2) to a level clearly suprathreshold and

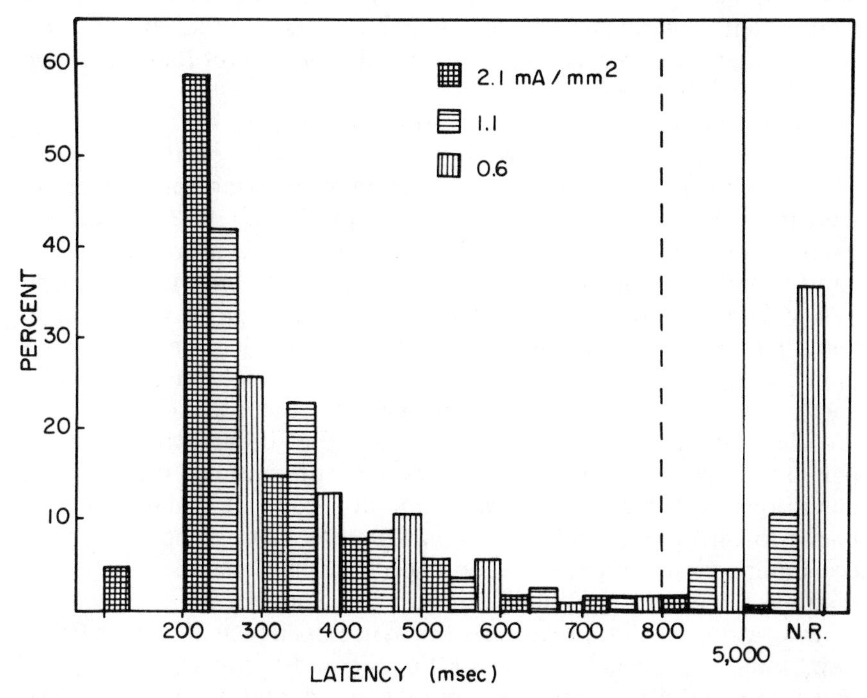

Fig. 2. Histograms of trial durations for group of 5 macaque monkeys receiving electrocutaneous stimulation of left calf at 3 intensities presented in sequence: 1.1 mA/mm^2, 0.6, 1.1, 2.1, 1.1, 0.6, 1.1, etc., for 21 trials per session. For both strong and weak stimuli, 95% of escape responses occur within 800 ms of shock onset. Thus, interval from 800 ms to 5 s (maximal trial duration) does not contribute significantly to escape-latency distributions. Decreasing stimulus intensity reduces proportion of early responses and increases percentage of trials not terminated by animals.

strongly painful (2.1 mA/mm^2; 3). Comparing the latency histograms for weak and strong stimulations: *1*) near-threshold stimulation produced few short trial durations (39% within 400 ms of trial onset), compared with 79% within the first 400 ms of trials involving strong stimulation. *2*) Approximately equal response percentages to the weak and strong stimuli occurred within the intermediate range of 400–800 ms, and the 800- to 5,000-ms period contained few responses to any intensity. *3*) The mild stimulus was tolerated on many more trials (36% nonresponses at 0.6 mA/mm^2 vs. 1% at 2.1 mA/mm^2). Thus, direct reduction of pain intensity from strong to weak does not extend the latency histogram, demonstrating that these animals decided whether to respond within 800 ms, regardless of the stimulus magnitude. Accordingly, when response latencies and timed-out trials are combined into a single plot of trial durations, the distribution of latencies is positively skewed, and for low intensities the distribution of durations is bimodal, with an unknown distance between the two peaks. That is, assignment of a value to the nonescape trials is arbitrary; these trial durations can simply be described as greater than the range of response latencies.

The escape-latency histograms presented in Figure 3 demonstrate the effects of a hypalgesic agent, morphine sulfate, on the timing of responses to stimuli evoking near-threshold, intermediate, or strong pain in normal subjects. At dosages ranging from 1 to 2 mg/kg, the extents of the distributions of latencies by macaques were not significantly lengthened beyond the control distributions, but the percentage of fast latencies was decreased and the percentage of nonresponse trials was increased by morphine. When the data from individual animals are scrutinized, two types of response pattern can be discerned. In Figure 4, histograms of responses to the low shock intensity are shown separately for three monkeys that differentiated escape percentage (group A) and two animals that differentiated the stimulus conditions on the basis of escape latency (group B). These different biases toward late responses or nonresponses were demonstrated by two independent variables: temporal order of each trial in daily testing sessions and administration of morphine (1–2 mg/kg im).

As a testing session progresses, the skin impedance decreases (from ~3.3 kΩ at the beginning to 2.3 kΩ at the end of the session; Fig. 5), and pain reactivity decreases for human and nonhuman primates (11). This permits a refined analysis of the relationship between pain-intensity and escape-latency histograms. Comparing groups A and B on early versus late trials (Fig. 4), some animals clearly varied the percentage of escape responses without altering the form of the distribution of latencies (group A), whereas the group B monkeys produced

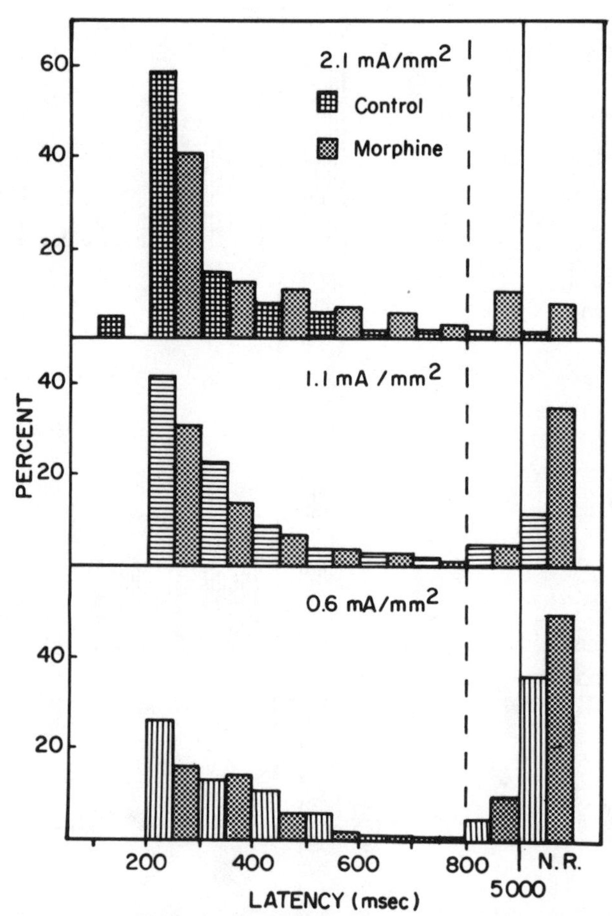

Fig. 3. Effects of morphine dosages (1–2 mg/kg) that are suprathreshold for significant effects on duration of shock tolerated by monkeys at current intensities normally evoking weak, intermediate, or strong pain sensations (0.6, 1.1, or 2.1 mA/mm^2, respectively). Morphine produces subtle shifts of latency distributions and percentage of escape responses.

longer latencies on late trials without increasing the percentage of nonescape trials. Similarly, morphine altered escape percentage for group A and escape latencies for group B. Thus, although each animal reveals reduced pain reactivity late in the sessions and after morphine administration, measurement of escape latencies or escape percentages would reflect the apparent hypalgesia only in ~50% of the animals. This result argues for a more general measure of response tendency that deals appropriately with bimodal distributions of trial durations and is conservatively sensitive to changes either in latency or response percentage.

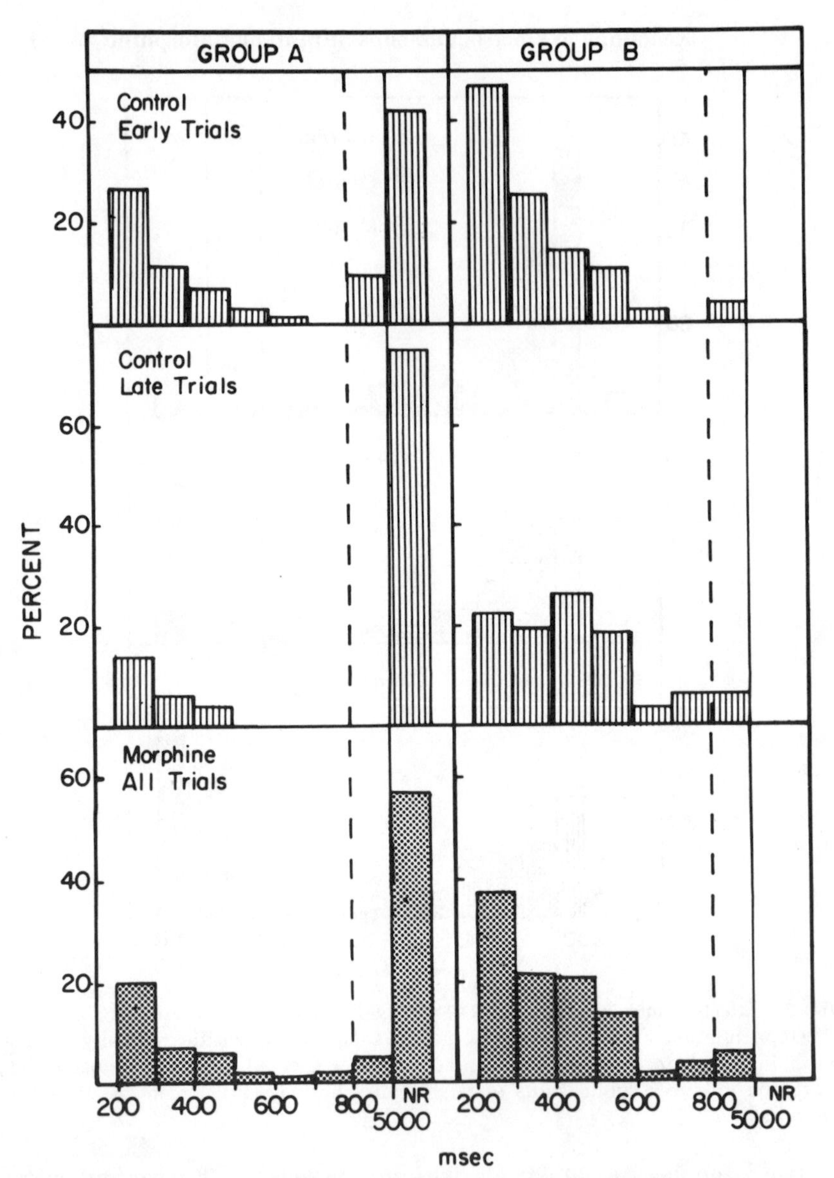

Fig. 4. Five macaque monkeys are divided into 2 groups that differentiate pain intensities and morphine versus control conditions based on changes in percentage of escape responses (group A; 3 animals) or of latency of escape responses (group B; 2 animals). All trials shown involve 0.6 mA/mm^2 stimulation, and trials 2 and 6 (early trials) or 14 and 18 (late trials), presented in 17 control sessions, are grouped to demonstrate diminished pain reactivity that accompanies repeated stimulation. Morphine (1–2 mg/kg) affects percentage of escape responses (group A) or latency of escape reactions (group B) in manner similar to influence of repeated stimulation. That is, when escape percentage provides more sensitive reflection of stimulus magnitude, this measure is more sensitive to morphine. Animals that differentiate stimulus intensities based on escape latency reveal presence of morphine by this measure.

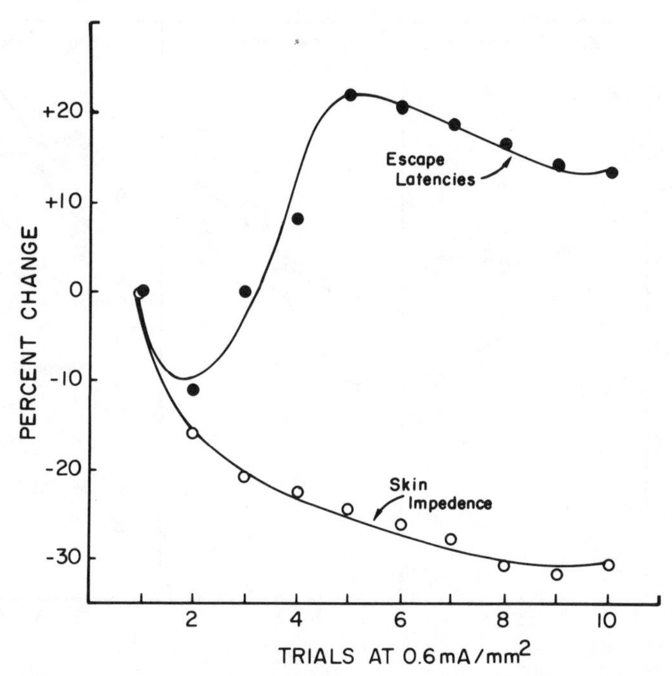

Fig. 5. Five macaques were tested for pain reactivity in 5 extended sessions involving 42 trials per session, with trial by trial monitoring of skin impedance. Both impedance and escape latencies for low-intensity trials (0.6 mA/mm^2), sequenced in saw-toothed pattern with 1.1 and 2.1 mA/mm^2, are shown as sessions progressed. Elevation of escape latencies over first half of session (trials 1–5 of 0.6 mA/mm^2) can be accounted for partly by a decrease in skin impedance occurring at highest rate in same period. Lower impedance requires lower voltage to produce constant current, and therefore stimulus power decreases as session progresses. A slight lowering of sensation magnitude is apparent to human subjects receiving stimulation in same pattern.

Figure 6 presents geometric means of latency distributions that are limited at the high end by restricting trial durations to a maximal value equal to the 95th cumulative percentile of response latencies on control trials at the low stimulus intensity. The intention is to let the animals dictate the range over which response latency varies. Having determined this range of meaningful variation of response speed, each distribution is adjusted to fit the normal response bias of the subject, in preference to the experimenter making an a priori choice of a maximal trial duration. The geometric mean is the appropriate measure of central tendency for the skewed distributions of escape latencies, and the separation of the modes is minimized. Maximal trial durations are conservatively estimated, but changes in escape latencies or in the percentage of escape responses are reflected by this measure of response tendency. An important advantage over restricted mea-

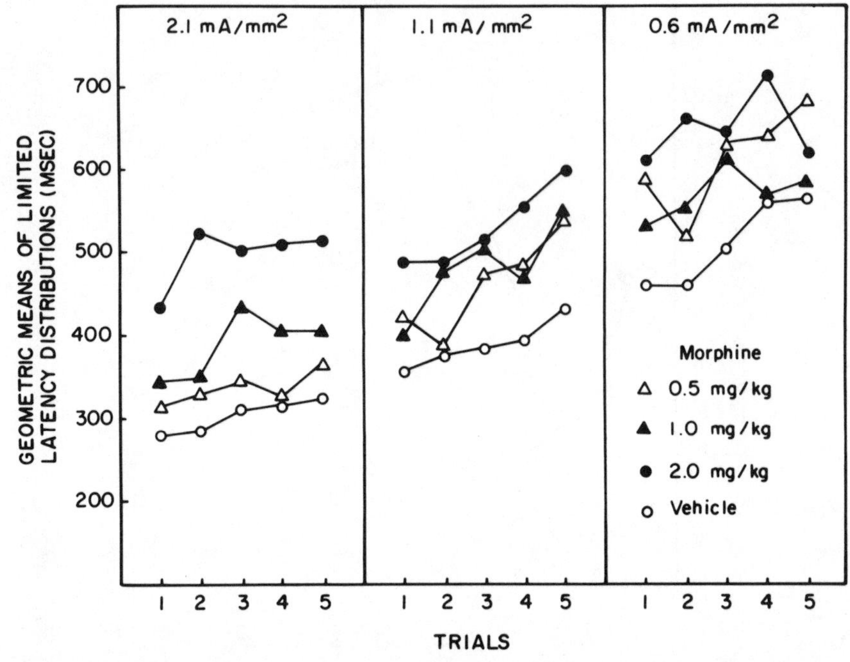

Fig. 6. Geometric means of restricted latency distributions are shown for successive trials at each of 3 intensities delivered in saw-toothed pattern during control sessions and sessions after 1 of 3 dosages of morphine sulphate. Data from stimulation at intermediate intensity are presented as successive pairs of trials (before and after each 2.1 mA/mm^2 trial). Trials effect is evident in control and morphine sessions. Differences in latencies between the 3 doses of morphine are significant only for strong shock. Doses of 1 and 2 mg/kg produce significant elevation of latencies for each stimulus intensity.

surement of escape latencies is that data from each trial are utilized, permitting analysis of effects on reactivity to stimulus intensities near pain threshold (where responses can be omitted, producing a variable pool of response latencies). An advantage of trial duration over percent escape as a measure of pain reactivity is that the latter measure plateaus near 100% at the intermediate levels of pain intensity. That is, percent escape is best at differentiating low from intermediate intensities, and latency most reliably distinguishes intermediate from strong levels of pain. When both measures of response tendency are combined by computing geometric means of constrained distributions of trial durations, a diminution in pain reactivity over successive trials at each intensity is revealed clearly for monkeys after injection of morphine or its vehicle (saline). Neither of the component measures (percent escape or response latency) reliably reveals the effect of successive trials that is obvious to human subjects and is shown in Figure 6.

Control for Response Suppression by Morphine

Apparently morphine decreases pain reactivity, as revealed by the geometric means of restricted response latencies, and it seems reasonable to assume that this effect results from a decrease in pain sensations elicited by each stimulus. This hypothesis is generated from the clinical use of morphine and by similar effects on many other behavioral assessments of pain reactivity. However, surprising as it may seem, the assumption that hypalgesia underlies the shift of response latencies to painful stimuli must still be questioned on a number of grounds (12, 13). *1*) The threshold for a significant effect of morphine on escape latencies for the macaques is 0.5–1.5 mg/kg (depending on the individual animal), and therapeutic dosages that are clearly suprathreshold for reduction of clinical pain in humans are typically an order of magnitude lower (0.05–0.15 mg/kg). *2*) Nearly all the laboratory animal studies of morphine hypalgesia have utilized high dosages of morphine (e.g., 2–12 mg/kg in monkeys and 5–10 mg/kg in rodents) that are required to depress behavioral measures that are questionably related to pain and highly sensitive to nonspecific behavioral suppression. *3*) Morphine produces substantial suppression of behavior, particularly at the dosages used in the experimental evaluation of hypalgesia, and yet controls are seldom employed for effects of morphine on motor reactions that are similar to the pain reactions but are elicited by nonpainful stimuli. Human subjects and patients tend toward inactivity and inattentiveness after morphine administration, and slowed responses are a natural consequence of the generalized behavioral suppression produced by morphine. The notion that opiates suppress pain in the context of a healing mode, involving withdrawal and inactivity rather than mobilization for adaptive reactions (chapt. nine), is suggested by the introspective reports of human subjects receiving therapeutic dosages of morphine in our experiments.

To compare the threshold and magnitude of nonspecific behavioral suppression with the effects of morphine on bar pulls to escape painful stimuli, five monkeys have been trained to perform the bar-pull response in the same apparatus to an auditory stimulus signaling the availability of food reinforcement. At dosages <2 mg/kg, food intake was not suppressed directly, and the animals appeared normally motivated for the food reinforcement. However, morphine at 1 mg/kg significantly elevated response latencies for food reinforcement (13). This effect of morphine on acquisitive response latencies demonstrates the need for caution in interpreting the effects of systemic morphine on the latency or percentage of responding to escape normally nociceptive stimuli. The generalized elevation of response latencies is particularly strong when the experimental paradigm promotes responses that are normally of long and variable latency. Most of the animal paradigms for algesimetry have restricted stimulus levels to a range below or near pain threshold for normal animals, and the response latencies are typically quite long. Long-latency terminations of gradually increasing stimulus intensities or of low levels of stimulation are highly labile, depending on levels of arousal and attentiveness, and they can often be construed as avoidance responses, manipulating future contingencies rather than escaping (reacting to) pain (13).

At dosages from 0.5 to 2.0 mg/kg, depending on the individual monkey's sensitivity to morphine, significant attenuation of operant escape reactions was detected by the measures of force and latency (3). As indicated above, the elevation of escape latencies was difficult to distinguish from a similar slowing of operant reactions to nonnoxious cues. Escape force was significantly decreased at similar threshold dosages (1–2 mg/kg), but the force of operant response for food reinforcement was not decreased significantly at these dosages (12, 13). Although encouraging in terms of an appropriate animal model for morphine hypalgesia, the threshold dosage for selective attenuation of the force of operant pain reactions is an order of magnitude above the dosages that appear to be quite effective in reducing clinical pain (0.05–0.2 mg/kg). An explanation offered for this discrepancy is that clinical pain depends on a coding process that is especially sensitive to opiates and is not ordinarily represented by reactions to phasically elicited pain (1). Much of the deep pain experienced by patients results from distension of viscera, ischemia, or destruction of tissue, with sensitization for chemical stimulation and prolonged discharge, which are characteristic of unmyelinated nociceptors (10). The unmyelinated nociceptors discharge tonically and at low rates, and C-fiber input to the dorsal horn of the spinal cord is selectively depressed by low dosages of opiates (6).

Importance of the Source of Experimental Pain

Acute experimental pain, whether elicited by thermal or electrical stimulation, is dominated ordinarily by sensations dependent on input from myelinated nociceptors. The fast pain sensation that most rapidly follows an incrementing thermal gradient or is first to onset after electrical stimulation is conveyed over myelinated nociceptors (2). Because electrical stimulation produces bursts of discharge in small myelinated (Aδ-) and unmyelinated (C) fibers that can be separated in time by stimulating distally and by varying stimulus intensity, the relative saliency and magnitudes of distinct first pain (from Aδ-fibers) and second pain (C fiber) sensations can be rated psychophysically (5).

When this is done, it is clear that a strong electrical stimulus produces an intense Aδ-related pain, followed by a weak delayed pain that wells up after the decline of the first sensation. Human subjects rated the first pain magnitude as 3–4 times that of the second pain resulting from stimulation at 2.1 mA/mm^2 (13).

After administration of 5 or 10 mg of morphine to normal human subjects (0.06–0.14 mg/kg), we have found that fast pain sensations from thermal or electrocutaneous stimulation are not decreased, but slow pain sensations from strong electrical stimulation are significantly diminished or lost (4). Similarly, patients receiving morphine for postoperative pain have reported that phasic pain from stretch or manipulation of an incision can be unaffected relative to chronic discomfort. Animals reacting to phasic electrical stimulation are responding to the dominant sensation attributable to activity in myelinated afferents, and thus it is understandable that the threshold dosage for attenuation by morphine is high. Measures of pain reactivity that are sensitive to systemic morphine only at dosages >0.5 mg/kg are probably revealing an effect on the intensity of pain conducted by myelinated nociceptors. However, controls for generalized behavioral suppression must be included within this range of dosages (the threshold for nonspecific effects will vary somewhat among individuals, species, and tasks). Also, the most powerful effects of morphine on pain are not assessed by these animal models. We must develop humane methods of independently evaluating reactions to myelinated and unmyelinated nociceptive inputs.

Summary

This chapter considers the reliability and validity of measures of operant speed and vigor for assessing the intensity of elicited pain in laboratory animals. Although subhuman species cannot make qualitative distinctions between pain sensations by verbal reports, it is feasible to assess the intensity of elicited pain by observing the intensity of behavioral reactions that eliminate painful stimulation. It is not sufficient to merely note the stimulus intensity at threshold for elicitation of escape reactions, because *1*) it cannot be proven that escape thresholds reflect pain thresholds, *2*) pain-detection thresholds can be quite insensitive to manipulation of suprathreshold pain intensities, and therefore *3*) it is important to describe changes in the form of the functions relating the speed or intensity of pain reactions to stimulus intensity. Control of strong pain is of greater clinical importance than modulation of weak pain, and thus it is necessary to evaluate reactions to stimuli at intensities above pain threshold. Furthermore, to claim that a measure reflects pain intensity, it is important to show that the behavior is affected differentially by suprathreshold levels of stimulation that evoke different magnitudes of pain.

The forms of stimulus-intensity, response-magnitude functions differ between animals, response types, and various experimental conditions, indicating that sensory magnitude cannot be inferred in absolute terms from these functions. However, relative changes in slope or displacement of stable base-line functions of individual animals as a result of experimental manipulations can represent strong evidence for increased or decreased sensitivity to the stimulus continuum. For example, the duration of electrical stimulation tolerated by monkeys is a response measure that is highly sensitive to several manipulations that clearly affect pain intensity. Because trial duration, response percentage, and response latency are commonly utilized as measures of pain reactivity, emphasis is placed on a method of combining these measures into a single estimate of response tendency that avoids problems inherent to each component measure. This procedure involves: *1*) determining the range of response latencies observed for each animal, *2*) setting the maximal trial duration at an upper limit of the latency distribution, and *3*) expressing the average trial duration as the geometric mean of the response latencies and the adjusted maximal durations. This measure avoids distortions of the component measures, adjusts to the response styles of individual animals, and is appropriately sensitive to manipulations of stimulus intensity and to decreases in pain sensation that occur with repeated stimulation during a testing session (Fig. 6).

Validation of algesimetric methods for laboratory animals depends on demonstrations of hyporeactivity or hyperreactivity to a variety of treatments that produce hypalgesia or hyperalgesia in humans. Furthermore, and equally important, the measures of pain reactivity should be selectively sensitive to manipulations of pain intensity. For example, a hypalgesic drug should attenuate pain reactions in proportion to dosages within an appropriate range and only when the behaviors are elicited by normally painful stimuli. Systemic administration of morphine to monkeys does produce a reliable attenuation of the operant measures that are most sensitive to direct manipulation of stimulus intensity or to decreases in the intensity of pain with repetition (Fig. 4). However, the effective dosages are an order of magnitude higher than utilized for chronic pain in humans, and the latencies of operant response for positive reinforcement are also elevated at these dosages. Thus, simple demonstration of increases in latency or decreases in the percentage of response to aversive stimulation cannot be taken as evidence of hypalgesia because of the powerful suppression of generalized operant behavior that is produced by morphine.

Companion evidence from human psychophysical studies indicates that the difficulty in demonstrating morphine hypalgesia lies with the stimulus rather than the response measures. Human subjects do not report reductions of the dominant onset sensations from painful electrical stimulation after injection of morphine at dosages that clearly reduce delayed (second) pain. Because the monkeys escape electrical stimulation within 2 s of onset, they are responding to first pain, and high dosages of morphine seem necessary to attenuate the onset sensation conducted over small myelinated afferents. The effects of morpine at 1–2 mg/kg on the operant escape measures, and the absence of a comparable reduction of the force of an acquisitive response, indicate that the operant measures are appropriately sensitive (or insensitive) to attenuation of phasic pain by morphine. Thus the challenge remains to develop humane methods of selectively stimulating unmyelinated nociceptive afferents so that the most powerful effects of morphine on nociception can be investigated.

Our research was supported by National Institute of Neurological and Communicative Disorders and Stroke Grant NS-07261 and National Institute of Mental Health Training Grant MH-15737. The animals involved were maintained in animal care facilities fully accredited by the American Association for Accreditation of Laboratory Animal Care.

We gratefully acknowledge the technical assistance of Jean Kaufman.

REFERENCES

1. Beecher, H. K. The measurement of pain. Prototype for the quantitative study of subjective responses. *Pharmacol. Rev.* 9: 59–209, 1957.
2. Collins, W.F., F. E. Nulsen, and C. T. Randt. Relation of peripheral nerve fiber size and sensation in man. *Arch. Neurol. Psychiatry* 3: 381–385, 1960.
3. Cooper, B. Y., and C. J. Vierck, Jr. A comparison of operant and reflexive measures of morphine analgesia. *Soc. Neurosci. Abstr.* 6: 430, 1980.
4. Cooper, B. Y., and C. J. Vierck, Jr. Morphine selectively reduces C fiber pain in humans. *Soc. Neurosci. Abstr.* 8: 510, 1982.
5. Hallin, R. G., and H. E. Torebjörk. Studies on cutaneous A and C fiber afferents, skin nerve blocks and perception. In: *Sensory Functions of the Skin in Primates*, edited by Y. Zotterman. Oxford, UK: Pergamon, 1976, vol. 27, p. 137–148. (Wenner-Gren Ctr. Int. Symp. Ser.)
6. Jurna, I., and G. Heinz. Differential effects of morphine and opioid analgesics on A and C fiber-evoked activity in ascending axons of the cat spinal cord. *Brain Res.* 171: 573–576, 1979.
7. Knibestol, M., and A. B. Vallbo. Intensity of sensation related to activity of slowly adapting mechanoreceptive units in the human hand. *J. Physiol. London* 300: 251–267, 1980.
8. Levine, J. D., D. T. Murphy, D. Seidenwurm, A. Cortex, and H. L. Fields. A study of the quantal (all-or-none) change in reflex latency produced by opiate analgesics. *Brain Res.* 201: 129–141, 1980.
9. Lineberry, C. G., and C. J. Vierck, Jr. Attenuation of pain reactivity by caudate nucleus stimulation in monkeys. *Brain Res.* 95: 110–134, 1975.

10. Perl, E. R., T. Kumazawa, B. Lynn, and P. Kenins. Sensitization of high threshold receptors with unmyelinated (C) afferent fibers. In: *Progress in Brain Research. Somatosensory and Visceral Receptor Mechanisms*, edited by A. Iggo and O. B. Ilyinsky. Amsterdam: Elsevier, 1976, vol. 43, p. 263–277.
11. Turskey, B., D. Greenblat, and D. O'Connell. Electrocutaneous threshold changes produced by electric shock. *Psychophysiology* 7: 490–498, 1971.
12. Vierck, C. J., Jr., and B. Y. Cooper. Suggested guidelines for assessing pain reactions in laboratory animal subjects. In: *Mechanisms of Pain*, edited by J. Liebeskind and L. Kruger. New York: Raven, in press.
13. Vierck, C. J., Jr., B. Y. Cooper, O. Franzen, L. A. Ritz, and J. D. Greenspan. Behavioral analysis of CNS pathways and transmitter systems involved in conduction and inhibition of pain sensations and reactions in primates. In: *Progress in Psychobiology and Physiological Psychology*, edited by J. Sprague and A. Epstein. New York: Academic, 1983, vol. 10, p. 113–165.
14. Vierck, C. J., Jr., O. Franzen, and B. Y. Cooper. Evaluation of electrocutaneous pain: (1) comparison of operant reactions of humans and monkeys and (2) magnitude estimation and verbal description by the human subjects. *Soc. Neurosci. Abstr.* 6: 430, 1980.
15. Vierck, C. J., Jr., D. M. Hamilton, and J. I. Thornby. Pain reactivity of monkeys after lesions to the dorsal and lateral columns of the spinal cord. *Exp. Brain Res.* 13: 140–158, 1971.
16. Vierck, C. J., Jr., C. G. Lineberry, P. K. Lee, and H. W. Calderwood. Prolonged hypalgesia following "acupuncture" in monkeys. *Life Sci.* 15: 1277–1289, 1974.
17. Vierck, C. J., Jr., and M. M. Luck. Loss and recovery of reactivity to noxious stimuli in monkeys with primary spinothalamic chordotomies, followed by secondary and tertiary lesions of other cord sectors. *Brain* 102: 233–248, 1979.

EIGHT

Concepts of General Anesthesia and Assessment of Adequacy of Anesthesia for Animal Surgery

Eugene P. Steffey

Department of Surgery, School of Veterinary Medicine, University of California, Davis, California

Pain is one of the most compelling and physiologically complex experiences of humans. It is usually a very unpleasant sensation that defies consistent satisfactory definition. What is commonly labeled as pain by humans covers a wide range of subjective phenomena, usually resulting from the perception of a noxious stimulus. Debate continues as to whether animals commonly used in teaching and research perceive pain as humans do (68).

Reactions to a noxious stimulus are classified in three group (5): *1*) skeletal muscle responses, *2*) reactions mediated by the autonomic nervous system, and *3*) responses dependent on pain perception. Pain perception is related to central nervous system (CNS) processing of incoming signals. It is currently believed that stimuli can be considered noxious or aversive to animals if they produce pain in humans, tissue damage (except perhaps electrical stimulation), and escape behavior in the animal (40, 68). It is universally recognized that stimuli associated with surgery meet these criteria, and humanitarian considerations require concern for insensitivity during surgery in experimental subjects.

The relief of surgical pain is primary justification for anesthesia. Consideration for the anesthetic management of experimental animals is of major importance to the design and results of biomedical research. Because anesthetics are not inert, the specific drug, the technique of administration, and the proper dosage must be carefully considered in light of animal comfort and survival as well as the effect of the agent on the experiment being conducted. I briefly focus on general concepts of pharmacologically induced general anesthesia used to minimize perception of surgically induced pain by experimental animals and methods for monitoring the adequacy of anesthesia. Specific anesthetic protocol is not described here. The majority of anesthetic techniques applicable to laboratory animals are similar to those used in veterinary and medical clinical practice.

Definitions

Anesthesia

Anesthesia is broadly defined as the loss of sensation in part or all of the body. It is most commonly induced by administering a drug or a combination of drugs, usually by parenteral injection or via the gastrointestinal or respiratory systems.

General anesthesia

General anesthesia provides overall insensitivity and unconsciousness. The level of insensitivity obtained during general anesthesia is usually related to drug dose. A surgical level of general anesthesia implies unconsciousness or sleep, amnesia, analgesia, and absence of motor response to noxious stimuli. An appropriate level of anesthesia may be produced by a single drug providing progressive loss of CNS responsiveness (e.g., hypnotic or inhalation anesthetic) or dissociation (e.g., ketamine) or by carefully selecting a number of drugs, each directed to a specific component of general anesthesia. The latter technique is clinically referred to as a method to produce balanced insensitivity and unresponsiveness or simply *a balanced technique*; the technique may include hypnotic, narcotic, inhalation anesthetic, and muscle-relaxant drugs.

There is no unequivocal best anesthetic technique for all animals and procedures. For example, minor surgery may be accomplished in awake or lightly sedated animals by injecting local-anesthetic drugs adjacent to the surgical site to interfere with pain transmission at a local level. On the opposite end of the spectrum, major surgical intervention may require intensive anesthetic management rivaling the utmost in contemporary medical sophistication.

Selecting Anesthetic Protocol

Important drug classes

In selecting a specific anesthetic protocol it is logical to consider the pharmacological requirements of the experiment, review the major drug classes, and select the specific drug(s) from the class(es) that best fulfills current needs. Six classes of drugs are of primary importance to this discussion: tranquilizer, hypnotic-sedative, narcotic-analgesic, dissociative, inhaled-anesthetic, and muscle-relaxant drugs. An in-depth survey of the characteristics of these drug classes as well as specific drugs within each class may be obtained from standard pharmacology or veterinary anesthesia texts (21, 26, 41, 56).

Tranquilizer. Tranquilizers, also referred to as ataractics or neuroleptics, are given to humans to relieve tension and anxiety. In most animal species these drugs do not readily produce sleep. Curiosity and aggression are usually decreased and mobility is reduced. In reasonable doses, drugs of this class do not generally affect spontaneous responses to conditioned stimuli or reflexes. Some tranquilizers also have potent autonomic nervous system blocking activity (e.g., α-adrenergic), antiemetic, and antihistaminic effects. Used alone, they are usually administered to animals to produce a calming effect and chemical restraint (i.e., tractability). They are sometimes given prior to or in conjunction with other centrally acting drugs (e.g., narcotics, hypnotic sedatives, inhalant anesthetics) to potentiate desired effects of the drugs while minimizing generalized depression of homeostatic mechanisms. Drugs frequently used in animal anesthesia include phenothiazines (chlorpromazine, promazine, acetylpromazine) and butyrophenones (droperidol). Benzodiazepines (diazepam) and thiazine derivatives (xylazine) also share some characteristics of the hypnotic-sedative drug class. Accordingly their specific classification may vary with different authors.

Hypnotic-sedative. The hypnotic-sedative class includes drugs that produce a spectrum of CNS-depressant effects (depending on dose) including sedation, sleep, general anesthesia, and (with overdose) coma and death. In low doses they are also anticonvulsant and produce relatively minimal direct autonomic depression. Common examples include barbiturates (thiamylal, thiopental, methohexital, hexobarbital, pentobarbital), chloral hydrate (may include varying concentrations of magnesium sulfate), chloralose, and urethan. These drugs are probably the most widely used in biomedical research for producing restraint and anesthesia in animals.

Narcotic-analgesic. Narcotic agents are primarily used to relieve or minimize pain. Narcotics may also produce a calming effect and sleep in some species (e.g., dog), whereas comparable doses may result in restlessness or excitement in others (e.g., cat or pony). Current preparations worth considering include morphine, meperidine, fentanyl, and etorphine (M 99). An agent not subject to regulations under the Federal Controlled Substances Act (nonnarcotic analgesic) is butorphanol. Unlike drugs from most other classes, direct antagonists, like nalorphine and naloxone, are available to rapidly reverse narcotic actions.

Dissociative. Dissociative agents are nonnarcotic, nonbarbiturate drugs that produce a state of analgesia (somatic) and unconsciousness in most species. In humans the unconsciousness is described as neither sleep nor anesthesia but rather a cataleptic state in which the individual appears conscious but is mentally "disconnected" or dissociated from the immediate environment. Although animals of various species are affected differently, generally many ocular, oral, and swallowing reflexes persist, muscle tone is increased, and high dose may produce tremors and convulsions. The agent of current primary interest to biomedical research is ketamine.

Inhaled anesthetic. The use of inhaled anesthetics in biomedical research is gaining popularity. Improved safety of contemporary agents and increased familiarity with their unique delivery is largely responsible for the increased use of these agents with larger laboratory animals. The advantages of using potent inhaled agents such as diethyl ether, halothane, methoxyflurane, enflurane, and most recently isoflurane to produce general anesthesia include ease of control of anesthetic depth, ease of dose quantitation, and ease in controlling respiratory gas tensions. Specialized equipment necessary for appropriate agent delivery is usually necessary and may pose some disadvantage.

Muscle relaxant. Muscle relaxants are usually characterized by their predominantly central or peripheral actions. Centrally acting agents include guaiacol glyceryl ether (guaifenesin) and are infrequently used except to provide recumbency of large domestic species such as cattle and horses. Peripheral neuromuscular blocking agents such as succinylcholine (depolarizing type), and *d*-tubocurarine, gallamine, and pancuronium (nondepolarizing type) are occasionally employed with general anesthesia. It is important to stress that neuromuscular-blocking drugs lack analgesic effects. They are used merely to prevent muscular movement and therefore minimize excessive anesthetic depth (infra vida). Consequently, analgesia and unconsciousness must be produced by appropriate concurrently administered drugs. In addition, because satisfactory spontaneous respiration is impaired or ceases with generalized paralysis, ventilation must be controlled with mechanical devices.

Drug combinations. In many circumstances administration of a single drug is not suitable for the animal and/or the investigation. In such cases a combination of drugs may be warranted. Each drug is selected to satisfy a specific need, usually potentiating desired effects or minimizing undesirable actions of concurrently administered drugs.

Examples, in addition to the use of muscle relaxants (supra vida), include providing tractability and analgesia by an ataractic-narcotic combination (i.e., neuroleptanalgesia) or minimizing or overriding the muscle rigidity usually associated with dissociative anesthesia by concurrent administration of a tranquilizer. Drug combinations important to these considerations are occasionally marked in a single vial (e.g., fentanyl + droperidol available as Innovar-Vet).

Other considerations

Laboratory facilities are important in ultimately defining the best anesthetic protocol. Restraining devices are necessary for some species and should be readily available (e.g., squeeze cage or restraining chair for primates and stocks or chute for large domestic animals).

If inhalation anesthesia is chosen for major surgery, a functional delivery apparatus appropriate to the selected species must be readily available. An inhalation box housing the entire animal during induction of inhalation anesthetic is most desirable for some species. Thoracotomy; the use of neuromuscular blocking drugs; or the need to maintain arterial carbon dioxide tensions normal, constant, or both mandates controlled positive-pressure ventilation. This is facilitated by a suitable mechanical ventilator. Finally, adequate monitoring equipment appropriate to the species and the investigation is obviously necessary.

Defining Adequate Anesthetic Level

The depth of general anesthesia must be defined to avoid the possibility that the animal perceives noxious stimuli as well as to minimize undesirable functional depression of life-support organ systems associated with greater-than-necessary doses of anesthetic drugs. For effective surgical anesthesia in animals, many investigators either administer a specific drug dose or seek specific signs indicative of adequate CNS depression.

Drug dose

Drug dose is predetermined by the investigator or taken from the published work of others. The dose relates to the potency of the drug as an anesthetic. Anesthetic potency is frequently described as the drug dose that eliminates purposeful movement (or some other response in which a reduction or loss of perception is implied) in response to a noxious stimulus in 50% of the animals studied (18, 28, 35, 51). Methodology usually includes administration of a known dose of anesthetic, application of a noxious stimulus, and observation of the kinds of reaction to the noxious stimulus. The dose of anesthetic may be expressed in terms of body mass or surface area of the animal. In the case of inhaled anesthetics the concentration or partial pressure of the agent in the end-expired (alveolar) gas is a more likely expression. It is most desirable if a drug-effected steady state is present—a notable advantage with inhaled anesthetics. Stimuli frequently used to determine effective anesthetic doses in larger laboratory animals such as dogs and cats include toe, paw, or tail pinch (applied with large forceps clamped to full lock across the paw or at the tail base); application of subcutaneous electrical currents; or surgical skin incision (18, 28, 51). As noted earlier, there are various reactions to noxious stimulation; i.e., skeletal muscle responses, reactions mediated by the autonomic nervous system, and most importantly pain perception—the processing by the CNS of the original stimulation. Purposeful animal movement or lack of movement in response to a stimulus is frequently selected as a reasonable end point for anesthesia. Indeed during the past 2 decades it has become one of the most common indices in anesthetic literature describing anesthetic potency. Although it is the result of activity in multiple areas of the body (including receptors, peripheral afferent fibers, synapses, interneurons, motoneurons and their peripheral fibers, motor end plates, and muscles) that may all be influenced directly by anesthetics (11), no other response characterizes the transition from no anesthesia to anesthesia so uniformly and consistently (18, 51).

In recent years the assessment of animal movement has been widely applied to the study and use of inhalation anesthetics. In 1963 Merkel and Eger (46) defined what has become a standard index of potency for inhalation anesthetics: MAC—the minimum alveolar concentration of anesthetic required to keep an animal from responding by gross purposeful muscular movement in response to a noxious stimulus. Information on the potency (MAC) of various inhalation anesthetics in different species is available elsewhere (16, 18, 51, 58, 62, 64–66).

The primary difficulty with administering a predetermined dose of anesthetic is that it does not account for responses peculiar to each animal. Therefore, depending on the dose selected for administration, some animals may react to surgical stimulation whereas others will be depressed to a degree in excess of that required or indeed desired. Numerous factors contribute to this. The concept of "move or no move" defines only one level of CNS depression, probably corresponding to a "barely surgical level of anesthesia," and is usually determined to describe only the median of the population. The dose of anesthetic effective in 50% of the population (ED_{50}), although important to the pharmacologist in comparing drug potencies, offers only limited assurance of uniform pain-free animal surgery. Most desirous to the scientist contemplating animal surgery and to the clinician is the anesthetic dose that anesthetizes 100% of the animals (ED_{100}). Unfortunately the ED_{100} can not be defined statistically. As a compromise the ED_{95}, though less precise than the ED_{50}, can be readily computed. For humans the ED_{95} is up to 40% greater than the corresponding ED_{50} (MAC) for common inhalation anesthetics (12).

A second cause for underestimating anesthetic needs is the variance in stimulus intensity during surgery and perhaps from the technique used to define potency. Strong stimulation is produced by peritoneal and visceral traction and corneal and periosteal manipulation. Anesthetizing doses based on weaker stimuli may be inadequate for more involved surgical manipulation.

Drug dosage established for one species may not be appropriate for use in another species for several reasons. Sensitivities of target organs to a drug differ between species. For example, morphinelike drugs act like CNS depressants in dogs yet at similar doses may produce excitement and convulsions in horses. The subhuman primate appears more susceptible to the circulatory depressant effects of some inhaled anesthetics than the dog or human exposed to equipotent concentrations (37, 58, 60). The anesthetic potency of nitrous oxide in many commonly studied species is only ~50% of that in humans and does not provide sufficient CNS depression by itself for surgical intervention (51, 58, 64, 65).

Major species differences exist in the rate and pattern of drug metabolism of various anesthetics and may account in part for differing doses. Meperidine, for example, is metabolized in the dog at a rate of ~70%/h, whereas humans metabolize it at a rate of only ~17%/h. Metabolism in the rat and mouse is even higher than in the dog (6). Johnston et al. (33) found that concentrations of fluroxene necessary for surgical anesthesia were lethal in dogs, cats, and rabbits, but not in humans. The difference in animal and human toxicity was subsequently shown to be related to differing metabolic pathways of the drug (34).

Additional modifiers of anesthetic requirement were recently reviewed by Quasha and co-workers (51). Concurrently administered drugs that cause an increase in the release of CNS catecholamines (e.g., dextroamphetamine) and hyperthermia may increase requirements for anesthesia. Conversely age; hypothermia; pregnancy; hypoventilation and arterial carbon dioxide tensions >90 mmHg; hematocrit <10%; and drugs such as narcotics, local anesthetics, sedatives, and tranquilizers decrease requirements for anesthesia.

Finally, the time-dependent reduction in effective plasma concentration of drug and the absence of drug-effected steady state after an injection limits the predictable duration of surgical anesthesia by noninhalational anesthetic agents.

Clinical signs of anesthesia

A second approach to appropriate anesthetic control is to base a graded administration of anesthetic on prevailing clinical signs. The need to evaluate an animal's response to anesthetics is commonly recognized in clinical veterinary medicine. This experience could and should be applied to laboratory endeavors, because insufficient anesthesia possibly allows perception of noxious stimuli, whereas too much anesthesia carries the risk of increased morbidity and death. Unfortunately, there is no simple description of an animal's response to an anesthetic during surgery.

Promising objective measures of adequacy of anesthesia have centered on the electrical activity of the brain as recorded from implanted or scalp electrodes. The oscillations of electrical potential generated by the many neurons and modified by the resistance of the tissues through which they pass forms the basis of the electroencephalogram (EEG). A major advantage of this method of analysis is that it can be used to monitor anesthesia by injectable as well as inhalation agents. However, disadvantages continue to limit its immediate usefulness for animal surgery in all but the most sophisticated research laboratories: i.e., *1*) the necessary equipment may not be readily available; *2*) the amount and complexity of data in a conventional EEG recording are generally overwhelming; and *3*) although changes with a specific anesthetic agent in an individual species may follow a fairly well-defined sequence, there is a lack of uniform change between species and anesthetic drugs (9, 10, 16, 20, 27, 32, 36, 39, 45, 47, 48, 52, 70, 71).

Sensory evoked responses have also been investigated with the intent to characterize transition from no anesthesia to anesthesia (8, 9, 13–15, 67). Unfortunately again, the relative lack of knowledge of the meaning and origin of evoked electrical activity, the many variations of the evoked potentials recorded, the multifactorial situation of a surgical operation, and the cost of necessary equipment further limit the present usefulness of these techniques.

More than 50 years ago Guedel (22, 23) codified a system to describe the level of responsiveness of humans anesthetized with diethyl ether (Table 1; 41, 56). The traditional classification considers a progressive depression of a continuum of CNS function. Guedel built on earlier attempts by others to divide the state of anesthesia into distinct packages, each correlating with a particular set of reflexes (50, 55). The system considers four stages of anesthesia and subdivides the

Table 1

Classic description of biological response sequence to general anesthesia

A. Stage I
   1. Stage of analgesia, induction, voluntary excitement, or analgesia and amnesia
   2. Period from beginning of induction to loss of consciousness
   3. Voluntary resistance to restrain and anesthetic vapors
   4. In humans: loss or obtundation of pain sensation
      Mental faculties are controllable throughout this stage but are progressively depressed until unconsciousness
B. Stage II
   1. Stage of delirium, involuntary excitement, or uninhibited action
   2. Period from loss of consciousness to onset of automatic respiration
   3. In humans: dream stage
C. Stage III
   1. Surgical stage
   2. Period from onset of automatic respiration to respiratory arrest
   3. Subdivided into 4 planes of anesthesia
      a. Plane I: light surgical
      b. Plane II: moderate surgical
      c. Plane III: deep surgical
      d. Plane IV: excessive surgical
D. Stage IV
   1. Stage of respiratory paralysis or overdose
   2. Interval between respiratory and cardiac arrest

third stage into four strata. Specific signs were considered indicative of a particular anatomical level of CNS depression.

Although the signs proposed by Guedel may still be observed, and the concept remains important, deficiencies exist. Numerous physiological responses widely used by clinicians to evaluate patient responses to anesthesics are not included (e.g., changes in arterial blood pressure). Secondly, Guedel's classic model was derived from unmedicated (prior to anesthesia) human patients allowed to breathe spontaneously during diethyl ether anesthesia, circumstances that differ widely from present discussion. Significant variability among animal species and anesthetics complicate the evolution of a simple uniform system (Fig. 1; 7, 10, 16, 41, 56, 70–72). Finally, Guedel's scheme does not address the reality that varying magnitudes of stimuli intensify the many factors modifying the signs of anesthesia during surgery.

Current standards of clinical anesthetic practice in veterinary medicine encourage the knowledge and use of an initial anesthetic loading dose necessary to suppress purposeful movement in the surgical subject and anesthetic manipulation related to continual stimulus-response assessment. This approach applies to all species of laboratory animals subject to major surgical intervention and usually requires little in the way of special equipment. It is tailored to the individual

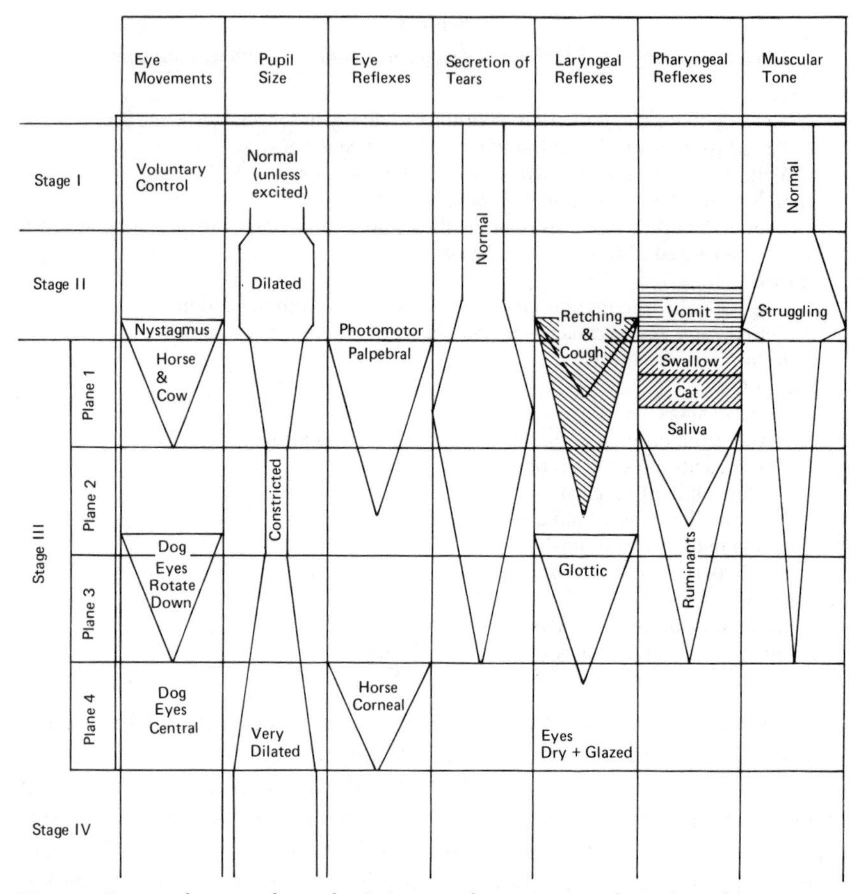

Fig. 1. Signs and stages of anesthesia in some domestic animals. [Adapted from Campbell and Lawson (7).]

animal and anesthetic agent(s). The major disadvantage of this approach is that it may require development of considerable skill in assessing signs that are unique to specific anesthetic drugs and/or species.

For simplicity, three levels of anesthesia are recognized: presurgical anesthesia, surgical anesthesia, and anesthetic overdose. The presurgical level or depth presumes perception of a noxious stimulus, is a level not acceptable for major surgical intervention, and in principle includes Guedel's first and probably second levels. Surgical anesthesia is further defined as light, moderate, or deep depending on the animal's response to the actual anesthetic dose and the level of stimulus intensity. For example, MAC defines the median alveolar dose of inhalation anesthetic that eliminates movement in response to a noxious stimulus. Evidence from human volunteers supports the notion that although skeletal movement may occur, analgesia and amnesia are present at concentrations of inhalation anesthetics less than MAC (1, 2, 44). Autonomic (i.e., adrenergic, cardiovascular) responses to surgical skin incision are abolished at anesthetic doses greater than MAC (53).

To limit accompanying unwanted depression and perhaps permanent damage of the animal's life-support systems, the majority clinical opinion encourages seeking the lightest possible level of anesthesia compatible with the surgical procedure.

Depth of anesthesia is estimated from changes in observed physiological variables indicated in Table 2 (10, 16, 41, 56). Some of these signs are more reliable than others as indicators of specific anesthetic

Table 2

Useful signs in clinical assessment of anesthetic depth

A. Cardiovascular system
   1. Heart rate and rhythm*
   2. Arterial blood pressure†
   3. Mucous membrane color
   4. Capillary refill time
B. Respiratory system
   1. Breathing frequency*
   2. Ventilatory volumes (tidal and minute ventilation)*
   3. Character of breathing*
   4. Arterial or end-tidal CO_2 partial pressure†
C. Eye
   1. Position and/or movement of eyeball†
   2. Pupil size*
   3. Pupil response to light
   4. Palpebral reflex
   5. Corneal reflex
   6. Lacrimation
D. Muscle
   1. Jaw or limb tone*
   2. Presence or absence of gross movement*
   3. Shivering or trembling*
E. Miscellaneous
   1. Body temperature
   2. Laryngeal reflex*
   3. Swallowing*
   4. Coughing*
   5. Vocalizing*
   6. Salivating
   7. Sweating
   8. Urine flow

Specificity in assessment of anesthetic depth for various animal species and anesthetic agents. * Moderate. † High.

depth. The actual value of each sign or measure is dependent on animal species and anesthetic agent.

Studies in animals of circulatory and respiratory effects of contemporary inhaled anesthetics conducted over the past few years have provided an opportunity to gather quantitative data relating signs of inhalation anesthesia to anesthetic dose (in multiples of MAC) in a variety of animal species. Results with inhaled anesthetics suggest that decreasing arterial blood pressure and increasing arterial or end-tidal (alveolar) carbon dioxide tension (during spontaneous ventilation) are signs of increasing anesthetic dose in healthy, surgically unstimulated dogs (Fig. 2), cats, calves, monkeys, ponies, and horses (46, 54, 60–63, 65, 66; unpublished observations). Although time may modify the absolute magnitude of some of these responses, relative dose-response changes persist (57, 59, 61, 63). Few quantitative studies are available for injectable anesthetics (43).

The intensity of stimulus applied to an anesthetized animal may rapidly and markedly alter the observed signs. A quiet animal with

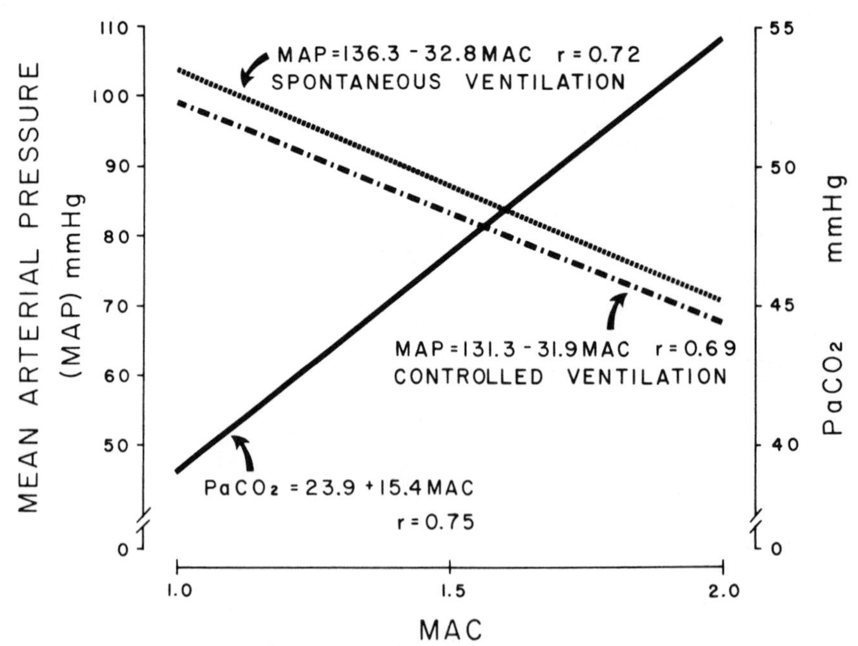

Fig. 2. Changes in mean arterial blood pressure (MAP) and arterial CO_2 partial pressure (Pa_{CO_2}) with increasing MAC multiples of halothane in O_2; MAC is minimal alveolar concentration of inhalation anesthetic preventing gross purposeful movement in response to noxious stimulus [tail clamp (8, 9)]. In dog, MAC for halothane in O_2 is 0.87%.

evidence of light-to-moderate anesthesia may quickly show evidence of light-to-insufficient anesthesia in the presence of noxious stimulation (Fig. 3; 17, 19, 25). Table 3 illustrates some general indications of inappropriate matches of anesthetic dose versus stimulus intensity. These are common responses averaged over many species and anesthetic agents because specific agent/species exceptions are countless.

Despite the natural tendency to rely solely on one or two observations, it is important to stress that no single sign of anesthesia is infallible. All possible signs should be observed in each animal, then with available information the best judgement possible can be made. A notable example is the use of movement as an exclusive sign of a

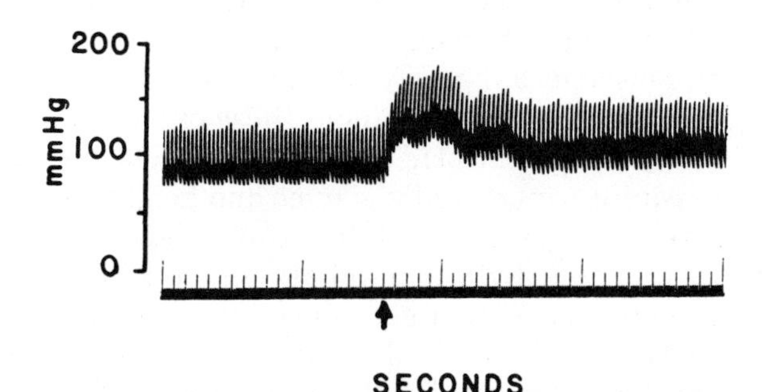

Fig. 3. Elevation of arterial blood pressure after skin incision in lightly anesthetized dog.

Table 3
Common responses to anesthetic dose-stimulus interaction

A. Signs of presurgical anesthesia
1. Bradycardia, tachycardia, arrhythmia
2. Arterial hypertension
3. Pupillary dilation, lacrimation, globe rotation
4. Tachypnea or breath holding
5. Deep breathing
6. Reduced alveolar/arterial P_{CO_2}
7. Limb/body movement
8. Salivating, vomiting
9. Swallowing
10. Laryngeal spasm
11. Phonating

B. Signs of deep surgical anesthesia
1. Bradycardia, tachycardia, arrhythmia, cardiac arrest
2. Arterial hypotension
3. Pupillary dilation, dry cornea, centrally fixed eye
4. Shallow breathing, respiratory arrest (*not* breath hold)
5. Elevated alveolar/arterial P_{CO_2}
6. Muscle flaccidity

positive reaction to a noxious stimulus. Muscle tone can be abolished at an anesthetic depth in excess of that necessary to produce amnesia, analgesia, and unconsciousness. However, this level of anesthesia has the disadvantage of concurrent depression of other vital organ functions. Alternatively, a neuromuscular blocking drug (NMBD) that selectively interferes with the synaptic transmission of signals from motor nerve to muscle cell may be (and frequently is) administered as part of the anesthetic protocol. This technique permits separate control of skeletal muscle contraction and affords a justifiable reduction in administered anesthetic and accompanying generalized depression. Unfortunately, to the uninitiated, use of NMBD eliminates many responses, including gross movement, normally relied on to determine level of CNS depression. Therefore they introduce the hazard that an animal may be conscious but paralyzed. Consciousness and awareness during surgery has been identified under these circumstances in some human patients (3, 4, 24, 29–31, 38, 42, 49, 54, 69). In an effort to prevent this possibility, increased vigilence and heightened awareness is necessary of remaining signs like relative and absolute arterial blood pressure, heart rate and rhythm, pupillary dilation, lacrimation, salivation, and in some species sweating. Objective measures of anesthetic dosage (e.g., delivered, inspired, alveolar, or arterial anesthetic concentrations) and EEG tracings also take on expanded importance in laboratories with suitable equipment and expertise.

Summary

Humanitarian considerations require concern for insensibility during surgery in experimental animals. Accordingly, anesthetics are administered. Unfortunately, an objective index does not exist that specifically and reliably correlates with pain perception. Consequently, clinical signs of anesthesia are used to attempt to define a stage of analgesia sufficient to minimize the likelihood of perceived cerebral recognition and recall. The majority of anesthetic techniques supportive of surgery on experimental animals are similar to those used in veterinary and medical clinical practice. Investigators unfamiliar with anesthetics and anesthetic protocols in animals and/or unskilled in the technical aspects of anesthesia are best advised to seek personalized outside professional assistance with regard to specific and perhaps unique anesthetic requirements.

REFERENCES

1. Adam, N. Effect of general anesthetics on memory functions in man. *J. Comp. Physiol. Psychol.* 83: 294–305, 1973.

2. Adam, N., and G. I. Collins. Alteration by enflurane of electrophysiologic correlates of search in short-term memory. *Anesthesiology* 50: 93–97, 1979.
3. Bahl, C. P., and S. Wadwa. Consciousness during apparent surgical anaesthesia. *Br. J. Anaesth.* 40: 289–291, 1968.
4. Blacher, R. S. On awakening paralyzed during surgery: a syndrome of traumatic neurosis. *J. Am. Med. Assoc.* 234: 67–68, 1975.
5. Beecher, H. K. The measurement of pain. *Pharmacol. Rev.* 9: 59–209, 1957.
6. Burns, J. J. Application of metabolic and deposition studies in development and evaluation of drugs. In: *Fundamentals of Drug Metabolism and Drug Disposition*, edited by B. N. LaDu, H. G. Mandel, and E. L. Way. Baltimore, MD: Williams & Wilkins, 1971, p. 340–366.
7. Campbell, J. R., and D. D. Lawson. The signs and stages of anaesthesia in domestic animals. *Vet. Rec.* 70: 545–550, 1958.
8. Chapman, C. R., and C. Benedetti. Nitrous oxide effects on the cerebral evoked potential to pain: partial reversal with a narcotic antagonist. *Anesthesiology* 51: 135–138, 1979.
9. Clark, D. L., and B. S. Rosner. Neurophysiologic effects of general anesthesia. I. The electroencephalogram and sensory evoked responses in man. *Anesthesiology* 38: 564–582, 1973.
10. Cullen, D. J., E. I. Eger II, W. C. Stevens, N. T. Smith, T. H. Cromwell, B. F. Cullen, G. A. Gregory, S. H. Bahlman, W. M. Dolan, R. K. Stoelting, and H. E. Fourcade. Clinical signs of anesthesia. *Anesthesiology* 36: 21–36, 1972.
11. Darbinjan, T. M., V. B. Golovchinsky, and S. I. Plehotkina. The effects of anesthetics on reticular and cortical activity. *Anesthesiology* 34: 219–229, 1971.
12. De Jong, R. H., and E. I. Eger II. MAC expanded: AD_{50} and AD_{95} values of common inhalation anesthetics in man. *Anesthesiology* 42: 384–389, 1975.
13. De Jong, R. H., F. G. Freund, R. Robles, and K. I. Morikawa. Anesthetic potency determined by depression of synaptic transmission. *Anesthesiology* 29: 1139–1144, 1968.
14. Domino, E. F. Effects of preanesthetic and anesthetic drugs on visually evoked responses. *Anesthesiology* 28: 184–191, 1967.
15. Domino, E. F., G. Corssen, and R. B. Sweet. Effects of various general anesthetics on the visually evoked response in man. *Anesth. Analg. Cleveland* 42: 735–747, 1963.
16. Eger, E. I., II. Monitoring the depth of anesthesia. In: *Monitoring in Anesthesia*, edited by L. J. Saidman and N. T. Smith. New York: Wiley, 1978, p. 1–14.
17. Eger, E. I., II, W. M. Dolan, W. C. Stevens, R. D. Miller, and W. L. Way. Surgical stimulation antagonizes the respiratory depression produced by Forane. *Anesthesiology* 36: 544–549, 1972.
18. Eger, E. I., II, L. J. Saidman, and B. Brandstater. Minimum alveolar anesthetic concentration: a standard of anesthetic potency. *Anesthesiology* 26: 756–763, 1965.
19. France, C. J., M. H. Plumer, E. I. Eger II, and E. A. Wahrenbrock. Ventilatory effects of isoflurane (Forane) or halothane when combined with morphine, nitrous oxide and surgery. *Br. J. Anaesth.* 46: 117–120, 1974.
20. Galindo, A. Central nervous system. In: *Modern Inhalation Anesthetics*, edited by M. B. Chenoweth. New York: Springer-Verlag, 1972, p. 103–122.
21. Goodman, L. S., and A. Gilman (editors). *The Pharmacological Basis of Therapeutics*. New York: Macmillan, 1975.
22. Guedel, A. E. Third stage ether anesthesia: a sub-classification regarding the significance of the position and movements of the eyeball. *Am. J. Surg.* 24, Suppl.: 53–57, 1920.
23. Guedel, A. E. Stages of anesthesia and reclassification of the signs of anesthesia. *Anesth. Analg. Cleveland* 6: 157–162, 1927.

24. Guerra, F. Awareness under general anesthesia. In: *Emotional and Psychological Responses to Anesthesia and Surgery*, edited by F. Guerra and J. A. Aldrete. New York: Grune & Stratton, 1980, p. 1–8.
25. Gur, M., and R. I. Purple. Blood pressure response as an indication of the anesthetic level in ground squirrels. *J. Appl. Physiol.* 40: 977–981, 1976.
26. Hall, L. W. *Wright's Veterinary Anaesthesia and Analgesia*. London: Bailliere, Tindall, & Cox, 1971.
27. Hatch, R. C., R. B. Currie, and G. A. Crieve. Feline electroencephalograms and plasma thiopental concentrations associated with clinical stages of anesthesia. *Am. J. Vet. Res.* 31: 291–306, 1970.
28. Heavner, J. E. Animal models and methods in anesthetic research. In: *Methods of Animal Experimentation*, edited by W. I. Gay. New York: Academic, 1981. p. 313–357.
29. Heneghan, C., R. McAuliffe, and D. Thomas. Morbidity after outpatient anaesthesia. *Anaesthesia* 36: 4–9, 1981.
30. Hilgenberg, J. C. Intraoperative awareness during high-dose fentanyl-oxygen anesthesia. *Anesthesiology* 54: 341–344, 1981.
31. Hutchinson, R. Awareness during surgery. A study of its incidence. *Br. J. Anaesth.* 33: 463–469, 1961.
32. Joas, T. A., W. C. Stevens, and E. I. Eger II. Electroencephalographic seizure activity in dogs during anaesthesia: studies with Ethrane, fluroxene, halothane, chloroform, divinyl ether, diethyl ether, methoxyflurane, cyclopropane and Forane. *Br. J. Anaesth.* 43: 739–745, 1971.
33. Johnston, R. R., T. H. Cromwell, E. I. Eger II, D. Cullen, W. C. Stevens, and T. Joas. The toxicity of fluroxene in animals and man. *Anesthesiology* 38: 313–319, 1973.
34. Johnston, R. R., E. I. Eger II, W. C. Stevens, and P. F. White. Fluroxene toxicity in dogs: possible mechanisms. *Anesth. Analg. Cleveland* 53: 998–1003, 1974.
35. Kissin, I., T. McGee, and L. R. Smith. The indices of potency for intravenous anaesthetics. *Can. Anaesth. Soc. J.* 28: 585–590, 1981.
36. Klein, F. F., and D. A. Davis. The use of the time domain analyzed EEG in conjunction with cardiovascular parameters for monitoring anesthetic levels. *IEEE Trans. Biomed. Eng.* 28: 36–40, 1981.
37. Lees, M. H., J. Hill, A. J. Ochsner, and C. Thomas. Regional blood flow of the rhesus monkey during halothane anesthesia. *Anesth. Analg. Cleveland* 50: 270–281, 1971.
38. Levinson, B. W. States of awareness during general anaesthesia. *Br. J. Anaesth.* 37: 544–546, 1965.
39. Levy, W. J., H. M. Shapiro, G. Maruchak, and E. Meathe. Automated EEG processing for intraoperative monitoring: a comparison of techniques. *Anesthesiology* 53: 223–236, 1980.
40. Lineberry, C. G. Laboratory animals in pain research. In: *Methods of Animal Experimentation*, edited by W. I. Gay. New York: Academic, 1981, p. 237–311.
41. Lumb, W. V., and E. W. Jones. *Veterinary Anesthesia*. Philadelphia, PA: Lea & Febiger, 1972.
42. Mainzer, J. Awareness, muscle relaxants and balanced anaesthesia. *Can. Anaesth. Soc. J.* 26: 386–393, 1970.
43. Maynert, E. W. The usefulness of clinical signs for the comparison of intravenous anesthetics in dogs. *J. Pharmacol. Exp. Ther.* 128: 182–191, 1960.
44. Mazzia, V. D., and C. Randt. Amnesia and eye movements in first stage anesthesia. *Arch. Neurol.* 14: 522–525, 1966.
45. McEwen, J. A., G. B. Anderson, M. D. Low, and L. C. Jenkins. Monitoring the level of anesthesia by automatic analysis of spontaneous EEG activity. *IEEE Trans. Biomed. Eng.* 22: 229–305, 1975.

46. Merkel, G., and E. I. Eger II. A comparative study of halothane and halopropane anesthesia. Including method for determining equipotency. *Anesthesiology* 24: 346–357, 1963.
47. Mori, K., M. Kawamata, H. Mitani, Y. Yamazaki, and M. Fujita. A neurophysiologic study of ketamine anesthesia in the cat. *Anesthesiology* 35: 373–383, 1971.
48. Mori, K., W. D. Winters, and C. E. Spooner. Comparison of reticular and cochlear multiple unit activity with auditory evoked responses during various stages induced by anesthetic agents. II. *Electroencephalogr. Clin. Neurophysiol.* 24: 242–248, 1968.
49. Mummaneni, N., T. Rao, and A. Montoya. Awareness and recall with high dose fentanyl-oxygen anesthesia. *Anesth. Analg. Cleveland* 59: 948–949, 1980.
50. Plomley, F. Stages of anaesthesia. *Lancet* 1: 134, 1847.
51. Quasha, A. L., E. I. Eger II, and J. H. Tinker. Determination and application of MAC. *Anesthesiology* 53: 315–334, 1980.
52. Robson, J. G. Measurement of depth of anaesthesia. *Br. J. Anaesth.* 41: 785–788, 1969.
53. Roizen, M. F., R. W. Horrigan, and B. M. Frazer. Anesthetic doses blocking adrenergic (stress) and cardiovascular responses to incision—MAC-BAR. *Anesthesiology* 54: 390–398, 1981.
54. Sia, R. L. Consciousness during general anesthesia. *Anesth. Analg. Cleveland* 48: 363–366, 1969.
55. Snow, J. *On the Inhalation of the Vapour of Ether in Surgical Operations: Containing a Description of the Various Stages of Etherization, and a Statement of the Results of Nearly Eighty Operations in which Ether has been Employed in St. George and University College Hospitals.* London: Churchill, 1847, p. 1–15.
56. Soma, L. R. *Textbook of Veterinary Anesthesia.* Baltimore, MD: Williams & Wilkins, 1971.
57. Steffey, E. P., T. Farver, and M. Woliner. Lack of temporal canine cardiovascular stability during constant alveolar concentrations of halothane anesthesia. *Federation Proc.* 40: 632, 1981.
58. Steffey, E. P., J. R. Gillespie, J. D. Berry, E. I. Eger II, and E. S. Munson. Anesthetic potency (MAC) of nitrous oxide in the dog, cat and stump-tail monkey. *J. Appl. Physiol.* 36: 530–532, 1974.
59. Steffey, E. P., J. R. Gillespie, J. D. Berry, E. I. Eger II, and E. A. Rhode. Circulatory effects of halothane and halothane-nitrous oxide anesthesia in the dog: controlled ventilation. *Am. J. Vet. Res.* 35: 1289–1293, 1974.
60. Steffey, E. P., J. R. Gillespie, J. D. Berry, E. I. Eger II, and E. A. Rhode. Cardiovascular effects of halothane in the stump-tailed macaque during spontaneous and controlled ventilation. *Am. J. Vet. Res.* 35: 1315–1319, 1974.
61. Steffey, E. P., J. R. Gillespie, J. D. Berry, E. I. Eger II, and E. A. Rhode. Circulatory effects of halothane and halothane-nitrous oxide anesthesia in the dog: spontaneous ventilation. *Am. J. Vet. Res.* 36: 197–200, 1975.
62. Steffey, E. P., and D. Howland, Jr. Potency of enflurane in dogs: comparison with halothane and isoflurane. *Am. J. Vet. Res.* 39: 573–577, 1978.
63. Steffey, E. P., and D. Howland, Jr. Cardiovascular effects of halothane in the horse. *Am. J. Vet. Res.* 39: 611–615, 1978.
64. Steffey, E. P., and D. Howland, Jr. Potency of halothane-N_2O in the horse. *Am. J. Vet. Res.* 39: 1141–1146, 1978.
65. Steffey, E. P., and D. Howland, Jr. Halothane anesthesia in calves. *Am. J. Vet. Res.* 40: 372–376, 1979.
66. Steffey, E. P., D. Howland, Jr., S. Giri, and E. I. Eger II. Enflurane, halothane and isoflurane potency in horses. *Am. J. Vet. Res.* 38: 1037–1039, 1977.
67. Uhl, R. R., K. C. Squires, D. L. Bruce, and A. Starr. Effect of halothane anesthesia on the human cortical visual evoked response. *Anesthesiology* 53: 273–276, 1980.

68. Vierck, C. J., Jr. Extrapolations from the pain research literature to problems of adequate veterinary care. *J. Am. Vet. Med. Assoc.* 168: 510–513, 1976.
69. Waters, D. J. Factors causing awareness during surgery. *Br. J. Anaesth.* 40: 259–263, 1968.
70. Winters, W. D. Effects of drugs on the electrical activity of the brain: anesthetics. *Annu. Rev. Pharmacol. Toxicol.* 16: 413–426, 1976.
71. Winters, W. D., K. Mori, C. E. Spooner, and R. O. Bauer. The neurophysiology of anesthesia. *Anesthesiology* 28: 65–80, 1957.
72. Woodbridge, P. D., III. Changing concepts concerning depth of anesthesia. *Anesthesiology* 18: 536–550, 1957.

NINE

Perspectives on Phylogenetic Evolution of Pain Expression

Stephen G. Dennis

Departments of Anesthesiology and Psychiatry and Behavioral Sciences, University of Washington, Seattle, Washington

Ronald Melzack

Department of Psychology, McGill University, Montreal, Quebec, Canada

Of the many stimuli encountered by an animal in its environment, few are as likely to elicit action as those disrupting or threatening to disrupt the physical integrity of the body. Exposure to noxious or potentially lethal forces is virtually unavoidable for most animals. From the indifferent assaults of the inorganic world to the directed attacks of predator, parasite, and pathogen, an individual is subject to various unpredictable and uncontrollable stimuli that signal imminent danger to life and well-being. The ability to detect such stimuli and to take action to minimize their effects undoubtedly represents a major selection pressure in animal phylesis. Individuals who lack this ability, e.g., those with various forms of congenital insensitivity to intense stimuli (15), have greatly reduced life expectancies. This amply attests to the adaptive value of nociceptive sensorimotor systems.

This chapter discusses how the adaptive value of nociceptive systems might be manifested in the lives of animals, and thus suggests ways by which natural selection based on nociception might have shaped the emergence of present-day species and their behavioral repertoires. Note that this discussion is speculative. Unlike other chapters in this book, which deal with hard data drawn from reproducible experiments, ours concerns the behavior and behavioral states of organisms long since dead. Behavior, unlike structure, leaves no trace in the fossil record, and the little about behavior that can be inferred from structure is often ambiguous. Moreover, even if one assumes that the striking structural and behavioral hierarchies of present-day species somehow reflect or preserve the sequence of past evolution, one still faces the unresolvable issue of where phylogenetically old behavior (or structure) stops and phylogenetically new behavior (or structure) starts. Thus the reader should recognize and be prepared for the dilemma inherent in the study of the phylogeny of behavior, i.e., despite being one of the most potent forces shaping the evolution of life forms on this planet, the behavioral repertoires and mental states of ancestral species remain forever unknown and inaccessible.

Although this limitation precludes definitive conclusions regarding the phylogeny of behavior in general, there is scientific merit in discussing the emergence of certain specific kinds of behavior, in particular those that are displayed virtually universally across the animal kingdom. It is one of the axioms of biology that the more fundamental a process is to life, as demonstrated by its representation across a wide diversity of life forms, the more likely it is that the underlying determinants and mechanisms of that process are conserved across species and down through the generations. Thus, studying the process in one species should provide valid information not only about other existing species, but also about the origins of the process in ancestral forms. Although one must be mindful that even fundamental processes can be subject to significant selective variation over time, it is in the universal processes that phylogenetic analysis is likely to be on its firmest ground.

This chapter deals with one such universal process: the escape from or avoidance of forces that damage or threaten to damage the body or internal organs. Distinctive behavioral responses to intense stimulation are displayed by virtually all animal species down to the protozoan, suggesting that rudimentary painlike behavior emerged at a very low level of organismic complexity. Such painlike sensorimotor processes might well have been meaningless to the unsentient, uncomprehending organisms that first experienced them, but it was presumably from these stimulus-response reflexes that highly complex patterns of pain expression and experience were gradually assembled and selected. How this might have occurred is a major question, but even a partial answer is beyond the scope of this chapter. Thus, except to reiterate the fundamental nature of painlike processes throughout the animal kingdom, the remainder of this chapter deals primarily with the nature of pain expression in higher forms, specifically the vertebrates.

The organization and specific themes of this chapter depart somewhat from the traditional approaches to this topic taken by comparative psychologists and anatomists. We have chosen to focus on the functional aspects of pain sensation and expression in the lives of animals, and to use this framework to illustrate the kinds of selection pressures that have presumably shaped and guided the phylogenetic evolution of pain expression. The discussion begins with some functional definitions of pain, focusing on the potential importance of distinguishing between pain arising from actual tissue damage and pain arising from the threat of damage. This theme is reprised in a brief review of the available evidence on how pain is signaled in the vertebrate nervous system. We conclude with a discussion of a factor not often considered in the phylogeny of pain: the social consequences of pain expression. In the course of these discussions, three putative selection pressures are identified as relevant to pain evolution.

Functional Definitions of Pain—Trauma Versus Threat

Pain is commonly defined as the unpleasant sensation that arises when the skin or other tissues are subjected to stimuli of sufficient intensity to cause or threaten damage. The two stimulus classes—damage and the threat of damage—are often considered together because both are indeed sources of pain. However, when considering the physiological mechanisms, behavioral consequences, or evolutionary origins of pain sensation and expression, it may be important to distinguish between pain due to actual damage and pain due to the threat of damage. They are different stimuli with different origins and potential implications for the animal. It has been hypothesized that they are mediated and modulated differently by the nervous system (3, 4).

Overt damage to the skin or internal organs presents the animal with several biologically significant problems: e.g., loss of blood and other fluids, entry of toxins into the system, and infection by microorganisms. Although nearly all species, certainly the vertebrates, have elaborate mechanisms for coping with such events, these healing and restorative processes take time and, depending on the severity of the trauma, may draw heavily on the body's metabolic capacity. Clearly, any process or activity that *1*) competes for the body's reserves of energy and material, *2*) increases the probability that the existing trauma could be worsened, or *3*) exposes the already debilitated animal to stimuli or situations that heighten the risk of additional injury will tend to lessen the probability that the animal will recover from the initial injury. It is reasonable to suppose, therefore, that mechanisms exist to suppress activity that might aggravate existing injuries. Pain associated with overt tissue damage undoubtedly represents one such mechanism. By imposing a powerful and persistent signal on the nervous system, the nociceptive elements from the damaged region can alter the animal's behavioral and motivational states so that behavior conducive to healing and restorative processes is favored, whereas behavior that might exacerbate the trauma is suppressed. Changes in locomotor activity, sleep, grooming, feeding, drinking, and social contacts are among the many behavioral alterations that potentially could derive from a trauma-related, pain-signaling system. The actions of these pain signals could be direct (i.e., could provide an unconditioned stimulus that elicits specific shifts in the behavioral repertoire) or indirect (i.e., could provide the potential for reinforcement of any action that tends to reduce the intensity of the nociceptive message). This damage-related pain signal presumably persists until the potential for reinjury is sufficiently lessened (e.g., through healing) or until a situation arises that carries even greater life-threatening potential than the risk of reinjury. Thus pain associated with overt tissue damage is expressed in behavioral patterns that tend to favor healing and recuperative processes.

Pain that arises from the threat of tissue damage is associated with very different circumstances and has very different behavioral implications. Between the initial contact of a potentially damaging stimulus and the onset of tissue destruction, there is an interval during which action by the animal could prevent or greatly reduce the damage. The duration of this threat interval depends on the nature of the offending stimulus and on the latency of the animal's response. In some cases the interval may be too short to afford any reasonable chance of avoiding injury no matter what action is taken. In other cases, however, the interval may be sufficiently long that a message sent from the tissue to the central nervous system will trigger a response in time to prevent damage. The probability of successful avoidance of damage cannot be estimated a priori because it depends entirely on circumstances. However, even if that probability were always very low, it is still reasonable to assume that a threat-related, response-triggering system could have significant adaptive value in minimizing the extent of injury, thus greatly reducing the dangers associated with overt tissue damage. It stands to reason that such a threat-related, damage-minimizing system causes pain sensation. A signal with intense negative affect not only compels the animal to action but also provides an immediate indicator of the success or failure of the specific actions taken. Moreover, encounters with noxious stimuli are often accompanied by various nonnoxious tactile stimuli. The negative affect associated with the truly dangerous components of the stimulus complex provides an unambiguous signal that cannot be lost in the "noise" of other stimuli. Thus pain associated with threat of tissue damage is expressed in immediate and vigorous actions directed toward the minimization of damage done by the offending stimulus.

Because the above considerations apply to virtually all species in virtually all environments, the two processes described may represent twin selection pressures that have guided and shaped the emergence of nociceptive systems across phylogeny. Although it is common to consider them together (i.e., as indistinguishable generators of pain sensation), it is clear that this blending may obscure the dual and perhaps conflicting evolutionary origins of pain sensation and expression. The appropriate behavioral response to overt damage may be inactivity; pain arising from trauma should presumably promote such behavior. However, the appropriate behavioral response to threat may be vigorous activity; pain arising from threat should therefore promote this kind of activity. Thus the overt expression of pain sensation may actually be a combination of inherently contradictory processes and behavioral tendencies. Perhaps the essence of the evolution of pain expression lies in the need to preserve the adaptive value of each process while integrating them into an adaptive whole.

How Pain is Signaled—Multiple Ascending Pathways

The anatomical and physiological substrates of pain signaling in vertebrates have been investigated for many years. A thorough review of this field is beyond the scope of this chapter, but several reviews are available (3, 6, 18; chapt. three in this book).

Perhaps the most important issue in our discussion is that pain signals appear to be transmitted to the brain by multiple ascending spinal pathways. Some of these pathways show considerable variation among different species, whereas others appear to be phyletically constant (7, 8). The meaning of this difference is uncertain, and a number of interpretations have been proposed.

The pain-signaling systems that remain relatively constant among species include the spinoreticular system (SRS), the paleospinothalamic tract (pSTT), and the propriospinal system (PSS). Probably the most significant of these is the SRS, a group of tracts that originate from cells located in the spinal gray matter, usually in the deeper regions. The axons of this system project rostrally in the ventral and ventrolateral zones of the spinal cord white matter, some fibers ascending ipsilateral and some contralateral to their cell bodies. The targets of the SRS are various cell groups, known collectively as reticular formation, located in the core of the brain stem as far rostral as the diencephalon. The SRS is conspicuous in vertebrates at the level of fish, amphibians, and reptiles (5, 12), and the size of this system relative to overall brain size tends to remain constant across vertebrate species.

The pSTT is generally much smaller than the SRS. It is a pain-signaling system that links the spinal gray matter directly to the medial and intralaminar thalamic nuclei. The majority of pSTT cell bodies are located in the deep layers of the spinal gray matter, although some may be found in the superficial regions (cf. chapt. three). Because the target nuclei of the pSTT projection also tend to be the targets of a more massive projection from the reticular formation, suggesting a major spinoreticulothalamic relay system, the pSTT could perhaps be viewed as a small, specialized spinal projection system for bypassing the reticular formation. Like the SRS, the size of the pSTT relative to overall brain size remains generally constant across species.

The third pain-signaling system that is prominent across vertebrate phylogeny is a rather complicated, loosely organized group of propriospinal pathways. This system is composed of tracts that originate and terminate in the spinal gray matter, forming a multisynaptic system that ascends the cord in a series of segmental links that may cross and recross the midline numerous times. Many of these PSS neurons have their cell bodies in the deep regions of the spinal gray matter (10) and their axons may be found in all quadrants of the cord white matter (11).

For convenience, these three systems will be referred to as the medial group, primarily because they all tend to project into the core of the neuraxis: spinal gray matter, reticular formation, and medial thalamic regions. In contrast, there are three other pain-signaling systems that constitute the lateral group. These systems show considerable variability across species.

The classic pain tract in humans and other primates is the neospinothalamic tract (nSTT). Rising primarily from cells in the spinal dorsal horn, the axons of this tract cross the midline in the spinal cord and ascend in the ventral and ventrolateral quadrants, without intervening synapses, to the ventrobasal and posterior nuclei of the thalamus. Although it is a rather small tract even in the higher primates, the importance of the nSTT in pain signaling is well established. Below the level of primates, however, the nSTT is considerably less prominent, both anatomically and in its role in pain signaling.

In contrast, the spinocervical tract is a projection system that is well developed in the lower vertebrates but relatively minor in the primates. Like the nSTT, the cell bodies of this system tend to be located in the spinal dorsal horns. The axons, however, generally ascend in the dorsolateral quadrants ipsilateral to their cell bodies. The target of this projection is a group of cells located at the border between the spinal cord and brain stem known as the lateral cervical nucleus (LCN). The LCN is clearly identifiable in carnivores but is much less prominent in primates. In humans the LCN may be quite variable (16). A projection from the LCN across the midline to the ventrobasal and posterior nuclei of the thalamus partially overlaps the nSTT projection to these same nuclei.

A rather surprising ascending pathway for pain signals was reported by Uddenberg (17) and Angaut-Petit (1, 2) in the dorsal columns of the cat. Although the dorsal columns are composed mainly of nonnociceptive primary afferents, they also contain a significant number of axons from secondary neurons whose cell bodies are located in the dorsal horns (13, 14). A portion of these postsynaptic fibers apparently can carry pain-related information, at least in the cat. These cells project to the ipsilateral dorsal column nuclei and from there to the contralateral ventrobasal and posterior thalamic nuclei. The pain-signaling capacity of this system in other species remains to be demonstrated.

In summary, the projection of pain signals from the spinal cord to the brain occurs via multiple ascending systems, some well represented across vertebrate phylogeny, others showing marked species differences. How each of these systems functions in the generation of pain-expressive behavior is unknown. Perhaps more important, but equally unknown, is how these various systems interact in the course of adaptive pain processing. Several models have been proposed to explain the respective roles of lateral and medial pain-signaling systems in the generation of pain sensation and behavior.

Melzack and Casey (9) suggested that the lateral and medial systems contribute differentially to the psychological dimensions of pain experience. Thus the lateral systems tend to convey information regarding the precise location, qualitative nature, and topographic extent of the noxious stimulus and thereby contribute mainly to the sensory-discriminative dimension of pain. The medial systems, on the other hand, tend to contribute more to the motivational-affective dimension of pain experience, via inputs to reticular formation and medial thalamus and then to the limbic system. These dimensions are presumably combined in a cognitive-evaluative dimension that integrates the nociceptive message into the existing behavioral context. Pain-expressive behavior, ranging from purely reflexive to complex and highly structured, is presumably generated at all points along the ascending chain. Because these three psychological dimensions apply to all pain experience, this conceptual model suggests that both the medial and lateral ascending systems participate in all pain-signaling processes.

A somewhat different though compatible point of view was suggested by Dennis and Melzack (3, 4). This model focuses less on the psychological aspects of pain experience and more on the capacity of both the medial and lateral systems to act as information-processing systems, each with a somewhat different spectrum of preferred stimuli or stimulus patterns. It was suggested that the lateral systems may be tuned preferentially to the onset of sudden changes in noxious stimulation and thus may be related to the threat modality of pain expression. The nociceptive message is conveyed rapidly along the neuraxis, eventually reaching the neocortex. At all levels it is presumably integrated into ongoing behavioral processes and the appropriate escape behaviors are initiated. The medial systems, on the other hand, may be preferentially tuned to the persistent or tonic components of noxious stimulation and thus may be better suited to the mediation of pain signals related to existing tissue damage. The continuous activity in this system is presumed to impose a persistent signal on the reticular formation and medial thalamus, and ultimately on the limbic system, to effect global changes in the animal's behavioral repertoire.

In both of the above models the distinctive patterns of species variability and constancy can be understood. The activity of the lateral systems, whether viewed as a contributor to the sensory-discriminative dimension of pain experience or as a threat-related, phasic information-processing system, is highly dependent on the general anatomy and behavioral capacities of the particular species. Thus it might be expected that the neuroanatomy of the lateral systems would change markedly with overt changes in species morphology. In contrast, the activity of the medial systems, whether representing the motivational-affective dimension of pain experience or a damage-related, tonic information-processing system, is more related to fundamental behavioral states and thus might be mediated by neuroanatomical structures that are more highly conserved across species. This is speculative, of course. Whether these or other models prove to be accurate reflections of the underlying mechanisms of pain sensation and expression remains to be determined.

Why Pain is Expressed—Individual Versus Species Considerations

The two behavioral classes discussed so far literally qualify as pain expressions. The vigorous activity of an animal escaping a noxious stimulus and the relative lethargy of an animal recovering from wounds are both overt reflections of an internal pain state, although not necessarily the same state. The adaptive value of these behavioral patterns is clearly measured by how effectively they minimize the probability and consequences of tissue damage to the individual animal. Thus, in these cases, why pain is expressed seems obvious: it is a manifestation of the fundamental drive for self-preservation.

There is another class of pain expression for which the answer may be somewhat less obvious. Responses like grimacing and vocalization have no apparent adaptive value in terms of escaping noxious stimuli or promoting healing, yet they are among the most common expressions of pain. Why do they occur? It is conceivable, of course, that they are merely incidental reactions that have no adaptive value. For various reasons, however, it seems more likely that they are in fact an important part of the repertoire of pain expression in animals, but that their adaptive value is measured more in terms of species survival than individual survival. Although the social consequences of pain expression seldom receive as much attention as other areas of pain research, it is reasonable to assume that social factors are indeed a major selection pressure governing the evolution of pain-signaling processes.

Sociobiological studies have shown that the adaptive value of a behavioral pattern or repertoire cannot be measured solely in terms of its benefit to the individual. Some actions, e.g., the so-called altruistic behaviors, greatly increase the likelihood that an individual will be harmed or killed, yet such behaviors persist in the repertoires of many species. Their value is measured in the increased chance of survival of other individuals of the same species. Often the beneficiaries are close relatives, according to the kinship hypothesis, but this need not always be the case. There are a number of possible social effects of pain expression.

Some socially relevant pain expressions may be extensions of the threat-related behavior. Vocalization, for example, may serve to warn conspecifics of potential or imminent danger. A single screech of pain can alert an entire social group to danger. That screech may do the victim little or no good, but it may greatly enhance the survival of the larger group (i.e., the gene pool from which the victim derived). Similarly, violent gestures or facial expressions may act as specific warning or sign stimuli that elicit adaptive behavior in conspecifics, even though they have no apparent benefit to the individual in pain.

A more subtle example of social interactions that actually could benefit the individual in pain as well as the social group concerns the possible releasing effects of pain expressions on nursing or tending behavior. The whimpers and grimaces of a wounded individual may stimulate other individuals to groom, feed, defend, or otherwise care for the victim. This may promote healing and recuperative processes that ultimately restore the wounded individual to a useful position in the group. It seems reasonable to assume that such social interactions do occur in human society. What role they play in the lives of other species remains to be determined.

Summary

Three selection pressures have been identified as potentially relevant to the phylogenetic evolution of pain expression: *1*) the exigencies imposed by overt trauma; *2*) the demands imposed by the threat of injury; and *3*) the social consequences of pain experience and expression. Together these and perhaps other pressures have shaped and guided the emergence of complex neural processes that underlie pain. It seems safe to conclude that such pressures will be in force for some time to come.

REFERENCES

1. Angaut-Petit, D. The dorsal column system. I. Existence of long ascending postsynaptic fibers in the cat's fasciculus gracilis. *Exp. Brain Res.* 22: 457–470, 1975.
2. Angaut-Petit, D. The dorsal column system. II. Functional properties and bulbar relay of the postsynaptic fibers of the cat's fasciculus gracilis. *Exp. Brain Res.* 22: 471–493, 1975.
3. Dennis, S. G., and R. Melzack. Pain-signalling systems in the dorsal and ventral spinal cord. *Pain* 4: 97–132, 1977.
4. Dennis, S. G., and R. Melzack. Comparison of phasic and tonic pain in animals. In: *Advances in Pain Research and Therapy*, edited by J. J. Bonica, J. Liebeskind, and D. Albe-Fessard. New York: Raven, 1979, vol. 3, p. 747–760.
5. Ebbesson, S. O. E. Brain stem afferents from the spinal cord in a sample of reptilian and amphibian species. *Ann. NY Acad. Sci.* 167: 80–101, 1969.
6. Hammond, D. L., and T. L. Yaksh. Peripheral and central pathways in pain. *Pharmacol. Ther.* 14: 459–475, 1981.
7. Mehler, W. R. The mammalian "pain tract" in phylogeny (Abstract). *Anat. Rec.* 127: 332, 1957.
8. Mehler, W. R. Some neurological species differences—a posteriori. *Ann. NY Acad. Sci.* 167: 424–468, 1969.
9. Melzack, R., and K. L. Casey. Sensory, motivational, and central control determinants of pain: a new conceptual model. In: *The Skin Senses*, edited by D. Kenshalo. Springfield, IL: Thomas, 1968, p. 423–443.
10. Molenaar, I., and H. G. J. M. Kuypers. Identification of cells of origin of long fiber connections in the cat's spinal cord by means of the retrograde axonal horseradish peroxidase technique. *Neurosci. Lett.* 1: 193–197, 1975.
11. Nathan, P. W., and M. C. Smith. Fasciculi proprii of the spinal cord in man: review of present knowledge. *Brain* 82: 610–668, 1959.
12. Noback, C. R., and J. E. Schriver. Encephalization and the lemniscal systems during phylogeny. *Ann. NY Acad. Sci.* 167: 118–128, 1969.
13. Rustioni, A. Spinal neurons project to the dorsal column nuclei of rhesus monkeys. *Science* 196: 656–658, 1977.
14. Rustioni, A., and A. B. Kaufman. Identification of cells of origin of nonprimary afferents to the dorsal column nuclei of the cat. *Exp. Brain Res.* 27: 1–14, 1977.
15. Sternbach, R. A. *Pain: A Psychophysiological Analysis.* New York: Academic, 1968.
16. Truex, R. C., M. J. Taylor, M. Q. Smythe, and P. L. Gildenberg. The lateral cervical nucleus of cat, dog, and man. *J. Comp. Neurol.* 139: 93–104, 1970.
17. Uddenberg, N. Functional organization of long, second order afferents in the dorsal funiculus. *Exp. Brain Res.* 4: 377–382, 1968.
18. Yaksh, T. L., and D. L. Hammond. Peripheral and central substrates involved in the rostrad transmission of nociceptive information. *Pain* 13: 1–85, 1982.

TEN

Species Differences in Drug Disposition as Factors in Alleviation of Pain

Lloyd E. Davis
Clinical Pharmacology Studies Unit, College of Veterinary Medicine, University of Illinois, Urbana, Illinois

The veterinarian has an absolute ethical obligation to alleviate pain and suffering in his animal patients. This commitment stems from the purposes of the profession, which every veterinarian swears to espouse (5). The fulfillment of this obligation may often be difficult because the veterinarian is asked to care for the health of the entire animal kingdom, with the exception of human beings. The essence of veterinary pharmacology, as opposed to human medical pharmacology, lies in the species differences existing in the effects, actions, disposition, dosage, and other properties of drugs (6). Among mammalian species, there is little difference in the actions and effects of various drugs (10); the principal variables affecting intensity of pharmacological effects are drug absorption, distribution, excretion, and biotransformation.

After oral administration, a pharmaceutical product must undergo disintegration and dissolution, and the molecules traverse a series of biological membranes to enter the circulation. Within the blood plasma, drug molecules exist freely in solution or bound to serum proteins. The unbound molecules are free to diffuse from the circulating blood into tissues, excretory organs, receptor sites, and sites of biotransformation. This complex movement of drug molecules within the body is illustrated schematically in Figure 1. Differences in these various processes will influence both the drug concentration attained at the site of action and the duration of the pharmacological effect.

This chapter reviews some of the differences encountered among common domesticated animals that influence the disposition of analgesic drugs. These considerations affect the selection of drugs, dosage,

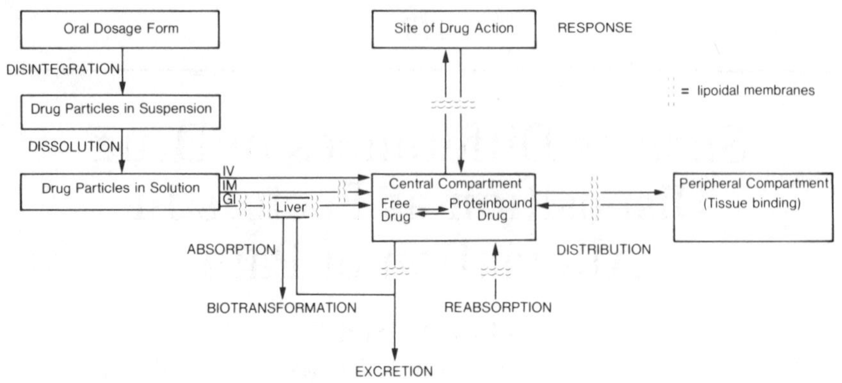

Fig. 1. Factors determining the disposition of drugs in the body. After administration, drug molecules traverse a series of lipoidal membranes.

and routes of administration for alleviating pain in domesticated animals.

Anatomical Features

The primary structural features affecting drug dispostion are differences in the alimentary tract, which adapt to accommodate dietary differences. The most common domesticated animals receiving veterinary attention are dogs, cats, swine, equines, and ruminants. The dogs and cats are carnivores with gastrointestinal tracts very similar to those of primates, except they are somewhat shorter in length. Swine are omnivorous and have a simple stomach, but they have a spiral colon. These features do not modify drug disposition to any extent, except perhaps in absorption of sustained-release products designed for human use. The equine species (horse, pony, ass, mule) are herbivorous, have a simple stomach, and possess a voluminous large colon and cecum (Fig. 2) that serve as sites for fermentive digestion of food. Under normal feeding conditions, the stomach is never empty (33). This tends to reduce the efficiency of absorption of orally administered drugs by equine animals.

The presence of the capacious rumen and reticulum markedly influences the disposition of orally administered drugs in ruminant animals (9). The forestomachs enable these marvelous creatures to convert cellulose and other components of forage to nutrients through fermentive processes. The rumen, reticulum, and omasum are serially interposed between the esophagus and the abomasum (i.e., true stomach) of ruminant animals (Fig. 3). The large capacity and the slightly acidic reaction of the ruminal contents provide a large potential "sink" into which basic drugs can sequester through ion trapping (24). The reticular groove (Fig. 4) is a structure interposed between the esophageal opening and the reticulum. The walls of the groove are muscular, allowing the groove to close into a tube form. This permits materials to bypass the rumen. The structures are lined by stratified squamous epithelium. Most drugs are poorly absorbed from the rumen, and some are destroyed by enzymes produced by the ruminal flora.

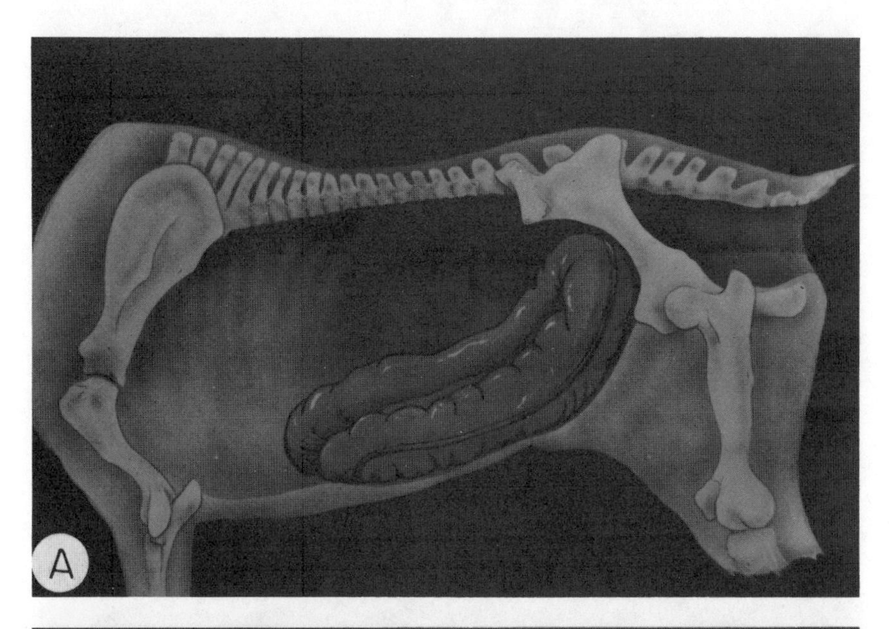

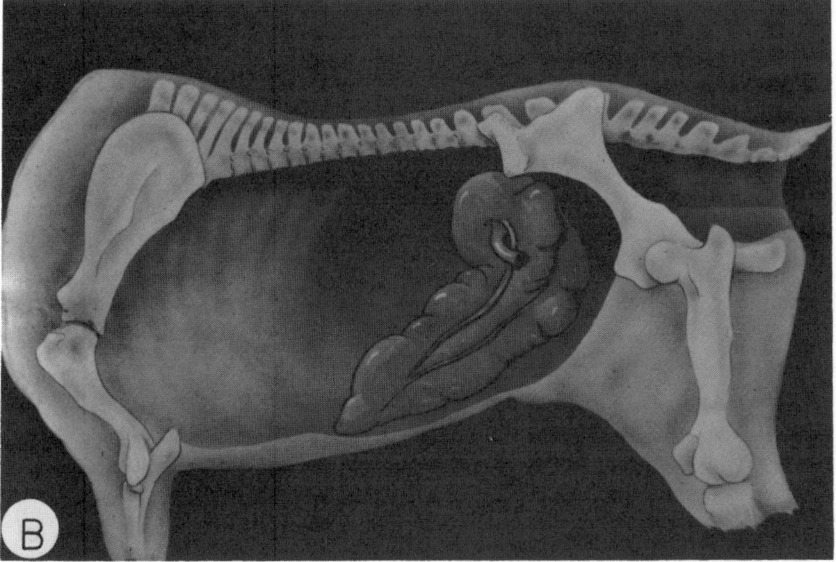

Fig. 2. Unique features of equine digestive tract. *A*: large colon. *B*: cecum.

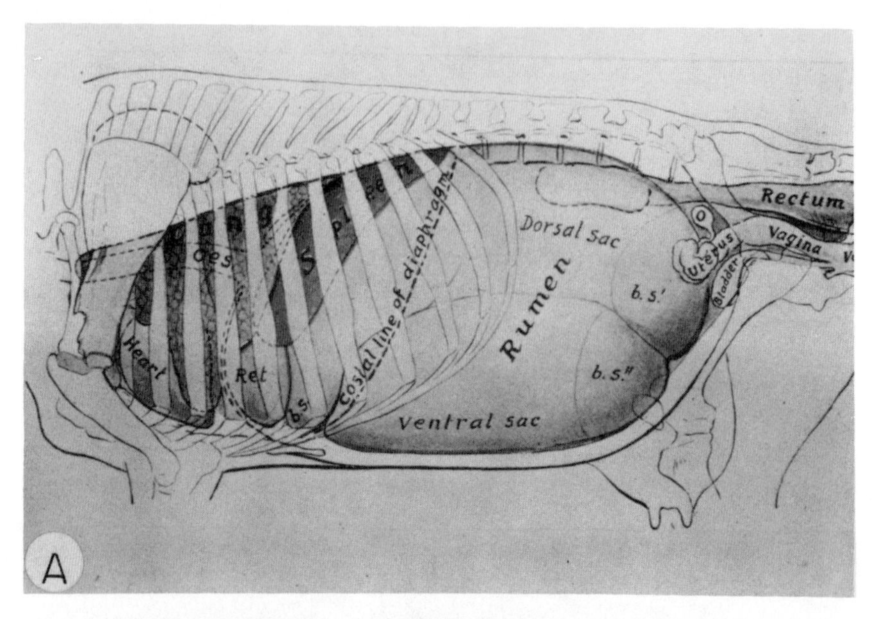

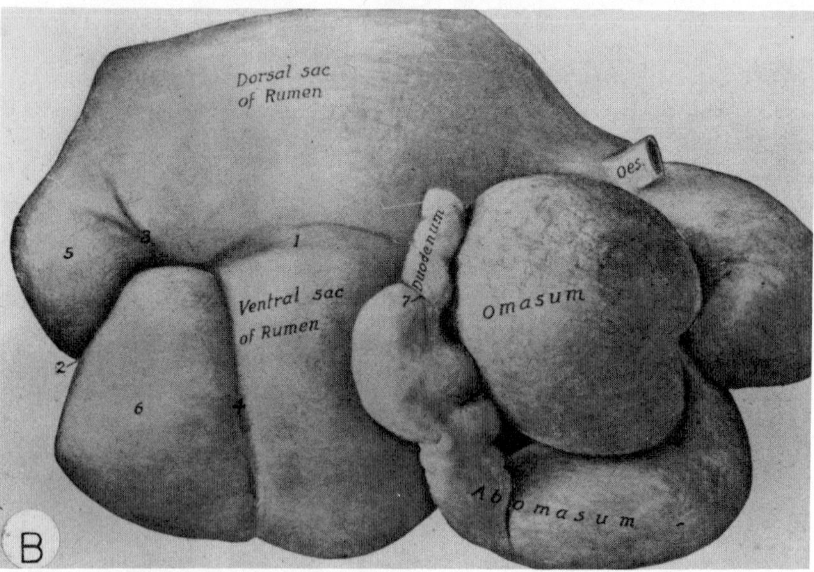

Fig. 3. Adaptations of stomach of ruminant animals. *A*: projection of viscera of cow on body wall, left view. *B*: forestomachs of ox, right view. The abomasum is homologous to true stomach. *C* (facing page): interior surface of rumen showing pillars and lining of stratified squamous epithelium. *D* (facing page): inner surface of reticulum. [Figs. 3*A* and 3*B* are from Getty (18a).]

Drug Absorption

The effects of these species differences in the structural components of the gastrointestinal tract on the absorption of certain analgesic drugs have been studied. Davis and Westfall (12) administered sodium salic-

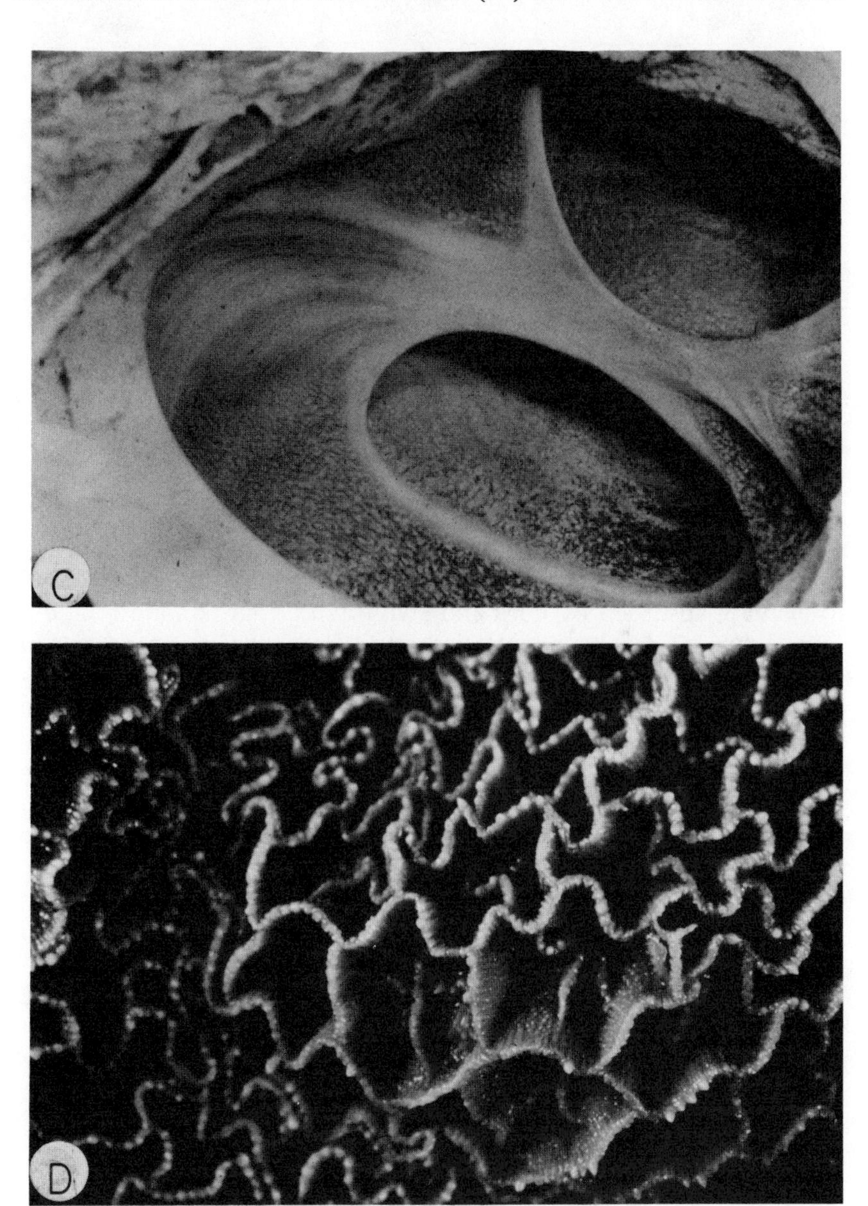

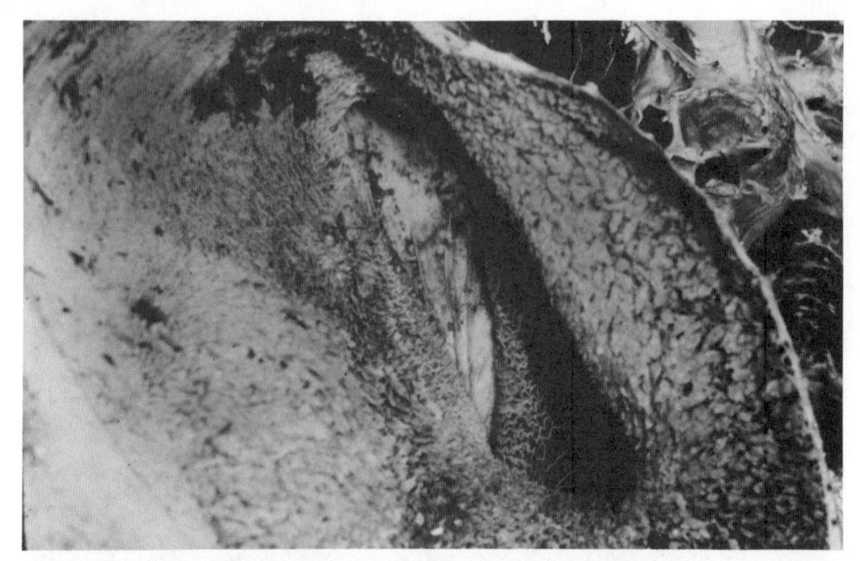

Fig. 4. Reticular groove of ox. Opening of esophagus is at top and groove extends into reticulum. View is from within rumen.

ylate orally to dogs, swine, ponies, and goats at doses of 120, 44, and 16 mg/kg. Blood samples were collected periodically and the salicylate concentrations were measured (Fig. 5). The drug was rapidly and extensively absorbed by dogs and swine with approximate relationships between drug concentrations and dosage. Concentrations of salicylate in the plasma of ponies were much lower but still somewhat dose related. Absorption from the rumen in the goat was slow, and thus resulted in negligible plasma concentrations of the drug. The secondary peaks were associated with drinking, which would cause ruminal liquor to pass into the abomasum. The slow absorption in goats was coupled with rapid elimination (see next section) from the plasma to produce very low plasma concentrations throughout the 10 h after administration.

Richez et al. (31) studied the effect of feeding on the absorption of lysine acetylsalicylate. Peak plasma concentrations of salicylate observed were 127 μg/ml at 69 min in fasting dogs, 104 μg/ml at 122 min in dogs fed dry food, and 88.7 μg/ml at 100 min in dogs fed a fatty meal.

Meclofenamic acid is another nonsteroidal anti-inflammatory drug with antipyretic and analgesic effects. It was studied in sheep and cattle by Marriner and Bogan (27), who investigated the effects of the rumen and the reticular groove on absorption of meclofenamate. Concentrations in ruminal, abomasal, and duodenal fluids and in plasma after oral administration of the drug (20 mg/kg) to sheep are shown in Figure 6. The change in plasma concentration was biphasic. The concentrations in ruminal and abomasal fluid indicate that a portion of the dose (administered as an aqueous suspension) was delivered directly through the closed reticular groove. Figure 7 shows the results of a comparable study in which the dose was administered directly into the rumen. A further experiment was performed comparing oral absorption of the drug in unweaned calves to weaned calves with functional rumen (Fig. 8). This provides further evidence for a portion of the drug passing through the closed reticular groove. The concentrations were much higher in the preruminant calves. Oral dosage was followed by biphasic absorption in weaned calves with the first peak corresponding to the single absorption peak seen in the unweaned calves.

Jenkins and Davis (25) studied the transfer rates of a number of

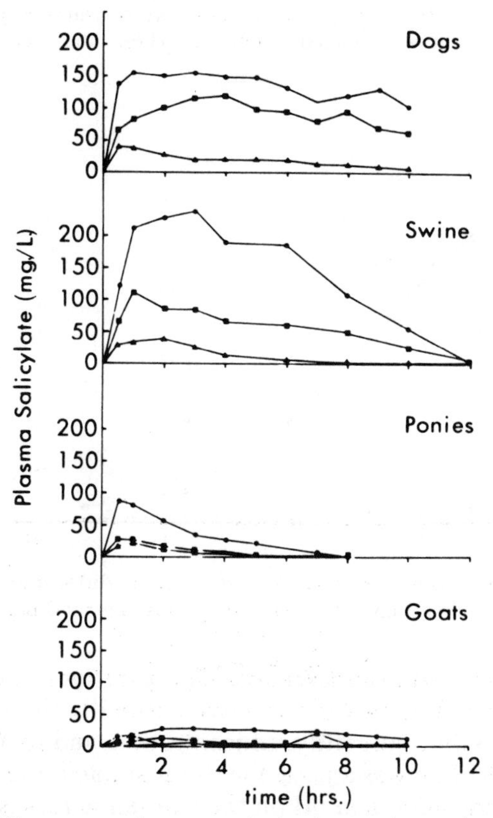

Fig. 5. Species differences in absorption of salicylate. Sodium salicylate (120, 44, and 16 mg/kg, as salicylate) was administered at time zero. [From Davis and Westfall (12).]

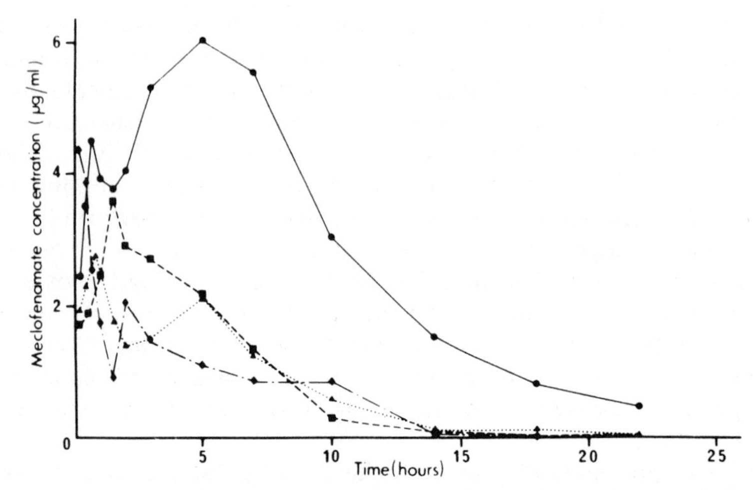

Fig. 6. Concentrations of meclofenamic acid in various fluids after oral administration (20 mg/kg) of the drug to sheep. Concentrations were measured in plasma (●), ruminal fluid (■), abomasal fluid (◆), and duodenal fluid (▲). [From Marriner and Bogan (27).]

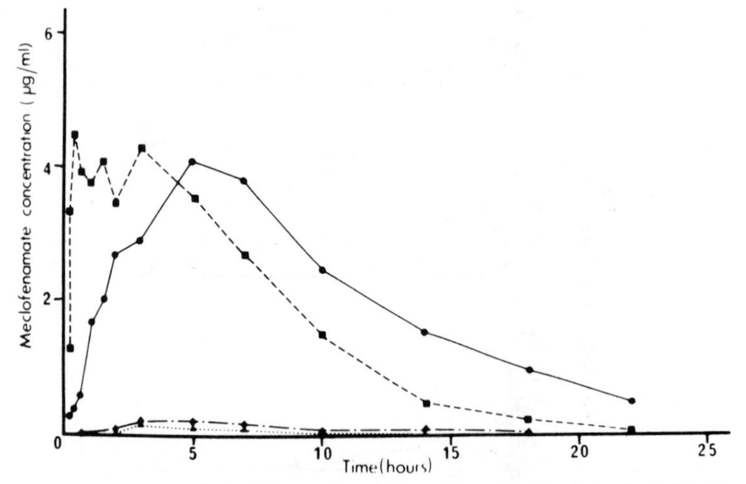

Fig. 7. Concentrations of meclofenamic acid in various fluids after administration into rumen of sheep. Symbols same as in Fig. 6. [From Marriner and Bogan (27).]

drugs across isolated ruminal mucosal preparations in vitro and across the intact ruminal epithelium in vivo. Generally the rates of transfer were related to lipid solubility, concentration, and ionization states of the drugs. Salicylate was among the drugs studied. Figure 9 shows the effect of concentration and pH on the transfer across isolated ruminal epithelium. The transfer of salicylate from the blood into the ruminal

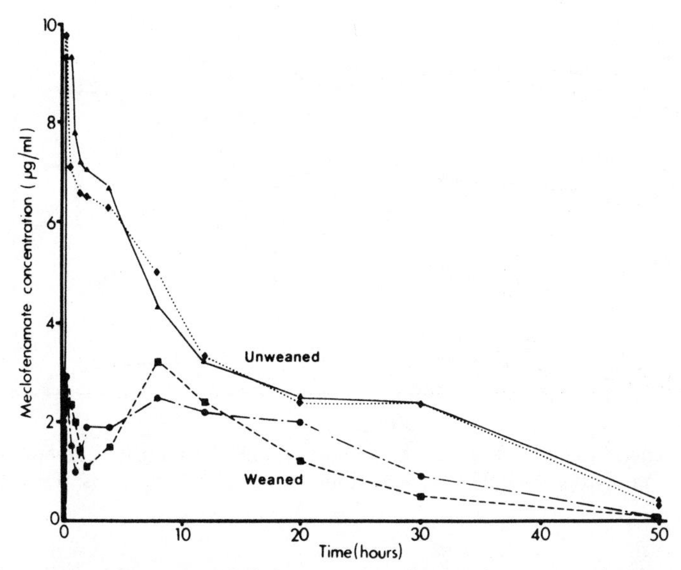

Fig. 8. Concentrations of meclofenamic acid in plasma of 2 unweaned calves (▲, ◆) and 2 weaned calves (●, ■) after oral administration of sodium meclofenamate (20 mg/kg). [From Marriner and Bogan (27).]

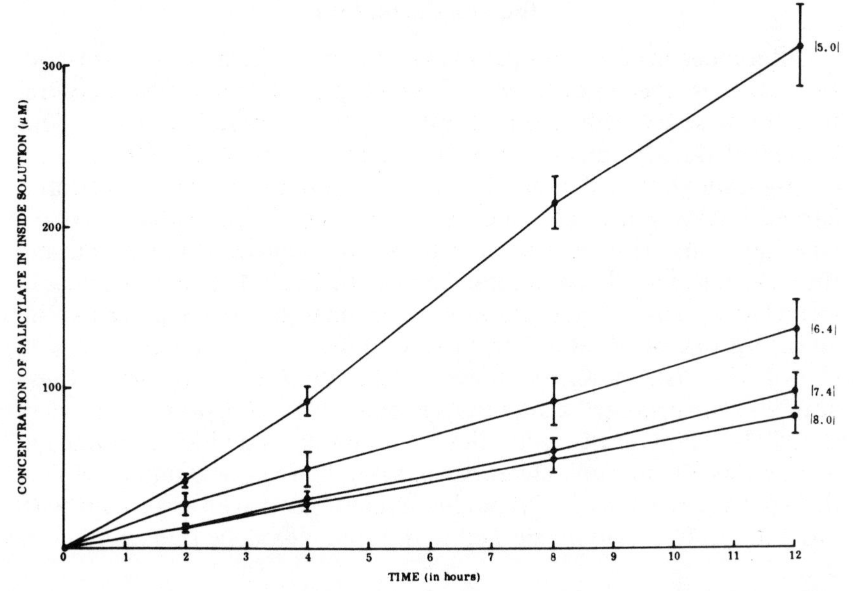

Fig. 9. Rate of transfer of salicylate across ruminal epithelium, from epithelial to serosal side in vitro. Concentration of salicylate in outside solution was 44 mM. The pH of outside solutions is in brackets. $n = 6$. [From Jenkins (24).]

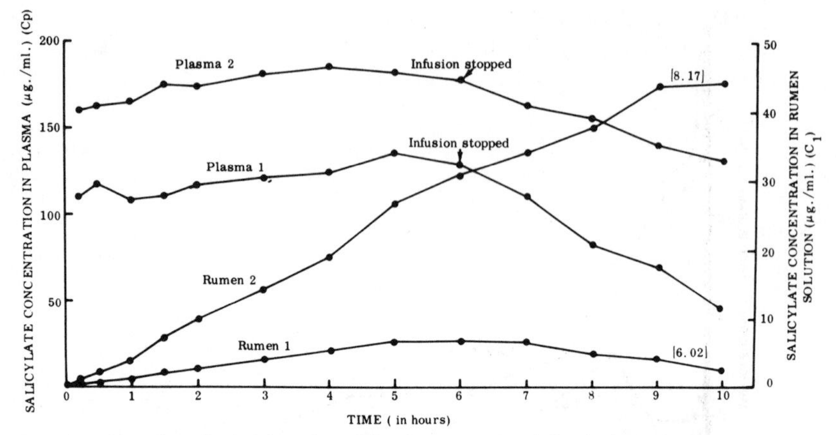

Fig. 10. Transfer of salicylate from blood into ruminal fluid, during constant intravenous infusion of drug. The pH of ruminal fluid is indicated in brackets. [From Jenkins (24).]

fluid, during constant-rate intravenous infusion, is shown in Figure 10. Note the effect of intraruminal pH on the drug concentrations appearing in the rumen.

Biotransformation

Differences in rates and pathways of biotransformation are responsible for most species differences in the time course of drug concentrations in plasma and hence in duration of therapeutic effect (21). This is very strikingly illustrated by two nonsteroidal anti-inflammatory drugs—salicylate and phenylbutazone. Sodium salicylate was administered intravenously at a dose of 44 mg/kg to goats, ponies, swine, dogs, and cats. The rates of elimination are shown in Figure 11, and the pharmacokinetic parameters appear in Table 1. The clearances of salicylate in ponies and goats were ~30 times greater than in cats. Not all of this difference was attributable to biotransformation. Figure 12 shows the distribution of salicyl metabolites in urine from these animals. The principal compounds formed in varying proportions were salicylate, salicylurate, and salicyl glucuronide. The high clearance in ponies was due to rapid clearance of salicylate in the alkaline urine of this species, whereas the high clearance in goats was associated with rapid biotransformation coupled with rapid clearance into an alkaline urine.

Species differences in the rate of elimination of phenylbutazone are shown in Table 2. This clearly demonstrates the futility and potential danger of extrapolation of data from one species to another with therapeutic intent. Initially horses were treated with phenylbutazone according to dosage regimens recommended for humans. The drug was considered to be worthless in horses until we learned about the

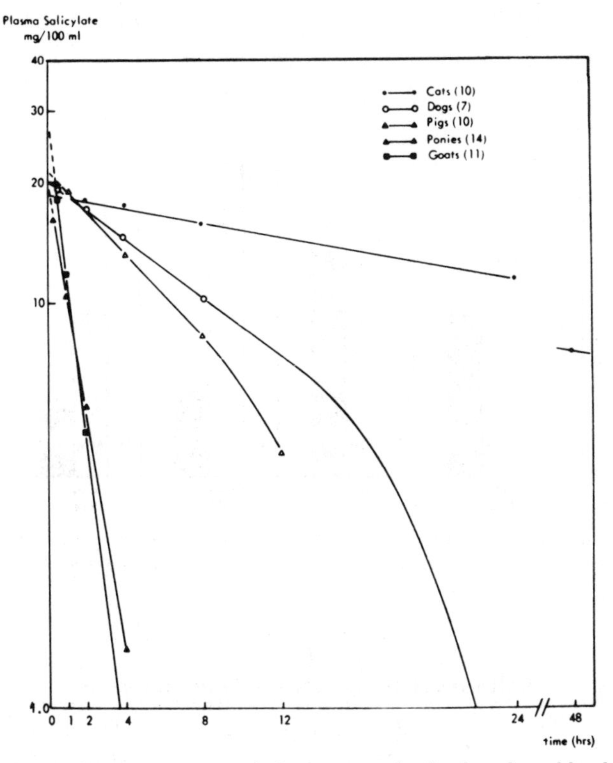

Fig. 11. Species differences in rate of elimination of salicylate from blood after intravenous administration of sodium salicylate. [From Davis and Westfall (12).]

Table 1
Pharmacokinetic constants for salicylate after intravenous administration of sodium salicylate (44 mg/kg)

Species	No.	B, mg/liter	β, h^{-1}	V_D, liter/kg	Cl_T, $ml \cdot kg^{-1} \cdot h^{-1}$	$t_{1/2}$, h
Cat	10	185	0.018	0.244	4.39	37.6
Dog	7	200	0.081	0.220	17.80	8.6
Swine	10	213	0.118	0.207	24.40	5.9
Pony	14	211	0.673	0.209	140.7	1.03
Goat	11	293	0.884	0.150	132.6	0.78

B, intercept of line describing elimination phase with ordinate; β, apparent first-order rate constant for elimination; V_D, apparent specific volume of distribution (a proportionality constant relating drug concentration in plasma to dose); and Cl_T, total-body clearance. [Adapted from L. E. Davis and B. A. Westfall (12).]

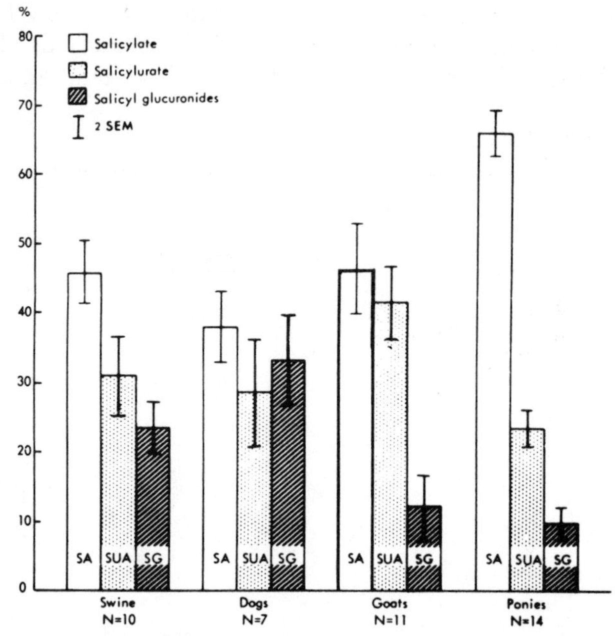

Fig. 12. Distribution of salicyl compounds in 24-h urine samples of several species after intravenous administration of sodium salicylate (44 mg/kg). [From Davis and Westfall (12).]

Table 2
Half-lives of phenylbutazone in several species

Species	$t_{1/2}$, h	Ref.
Human	72	13
Ox	55	14
	42	15
Goat		
Female	19	3
Male	14.5	
Cat	18	7
Rat	6	13
Dog	6	13
Swine	4	23
Baboon	5	13
Horse	3.5	30
Rabbit	3	13

pharmacokinetics of phenylbutazone in the horse. Phenylbutazone, unlike salicylate, is slowly eliminated from ruminants, and the rate in the goat is much greater than in the ox.

The opiates and opioids are the most valuable drugs for alleviating severe somatic or visceral pain. There has been considerable confusion regarding dosage of these drugs in the past because of lack of pharmacokinetic information. Sensitive methods for determining morphine concentrations were not available until recently (32). The extent of species differences in the elimination of morphine has not been established. Clinically the duration of analgesic response in the dog appears to be ~5–6 h. The elimination half-life in dogs has been determined to be 75 min (22) or 67 min (18), which seems inconsistent with the clinical observations. However, Hug et al. (22) found that elimination of morphine from cerebrospinal fluid (CSF) was slower ($t_{1/2}$ CSF = 123 min) than from blood. Sixty-six percent of the dose was recovered in the urine within 6 h after intravenous injection. Most of the morphine dose is excreted by the dog as morphine glucuronide. With a relatively high dose (1 mg/kg) and an insensitive-assay method, the half-life of morphine in the cat was estimated to be 3.05 h (8). Much of the past reputation of morphine in cats resulted from overdose. Several generations of veterinary students were taught that the use of opiates in cats was absolutely contraindicated because of the mania that high doses of morphine produces in them. At a dose of 0.1 mg/kg, morphine provides analgesia (without excitement) lasting 6–7 h (8). Similarly, dogs are generally given more morphine than is necessary for effective analgesia. A dose of 0.25 mg/kg of morphine hydrochloride administered subcutaneously provided apparent pain relief for 5–6 h in dogs after thoracotomy (L. E. Davis and P. E. Chase, personal observations).

Morphine is not generally used in horses, swine, or ruminant animals for pain relief. Swine and many horses are excited by morphine and may become dangerous to animal handlers. Morphine does not exert any apparent effect on ruminant animals. The reason for this is unclear. An interesting observation has been made in cattle (20). Morphine was found to be a common component in cow's milk and probably was derived from plant sources in the animal's diet.

Little pharmacokinetic information is available concerning methadone in domesticated animals. Clinically, it appears to have effects and duration of action similar to morphine in dog and cat. It is preferred to morphine for use in horses because it provides more reliable sedation.

Meperidine enjoys an undeserved reputation as an analgesic drug in veterinary medicine. Its half-life in the cat is 0.7 h because of rapid demethylation to normeperidine (8). Clinically, in both dog and cat, the duration of analgesia is ~45 min. The same problem occurs in horses, in which the brief duration of action is associated with a half-life in the body of 66 min (1).

We have studied pentazocine in domesticated animals (11). The rates of elimination were rapid in all of the species studied (Table 3). Clinical use of the drug in these species has corroborated our observations because the period of pain relief is short.

Oxymorphone is a morphine derivative that is ~10 times as potent as morphine. It has advantages over morphine or methadone for relieving pain in dogs, cats, or horses under certain circumstances. Oxymorphone's depressant effects on the central nervous system are slight (29), and it does not depress the cough center (26). This could be advantageous in some postsurgical situations when the animal must be able to keep its airway clear of fluids.

Two veterinary drugs used for their analgesic effects are flunixin meglumine and xylazine hydrochloride. Flunixin (Fig. 13) is a nonsteroidal anti-inflammatory drug that is about four times as potent as phenylbutazone. It is primarily administered to horses. The half-life of flunixin in horses was found to be 1.6 h (35). Xylazine is a sedative analgesic used in dogs, cats, swine, horses, cattle, and exotic animals (Fig. 14). This drug is exceptional because the species differences in effects of the drug cannot be attributed to variations in drug disposition. Garcia-Villar et al. (17) investigated the comparative pharmacokinetics of xylazine. Their data are compared with the duration of action and dosage of xylazine in different species in Table 4. There was some

Table 3
Pharmacokinetic parameters for pentazocine in domesticated animals

Species	B,* μg/ml	β,* min^{-1}	$t_{1/2}$,* h	V_D,* liter/kg	Cl_T, $ml \cdot kg^{-1} \cdot min^{-1}$
Pony	0.59	0.0071	97.1	5.09	36.0
Goat	0.52	0.0136	51.0	5.77	78.5
Swine	0.63	0.0143	48.6	4.76	68.0
Dog	0.85	0.0313	22.1	3.66	115.0
Cat	1.08	0.0083	83.6	2.78	23.0

* See Table 1 for definitions. [Modified from Davis and Sturm (11).]

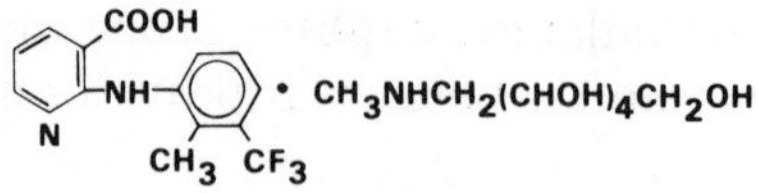

COOH NH · $CH_3NHCH_2(CHOH)_4CH_2OH$ N CH_3 CF_3

Fig. 13. Structure of flunixin meglumine.

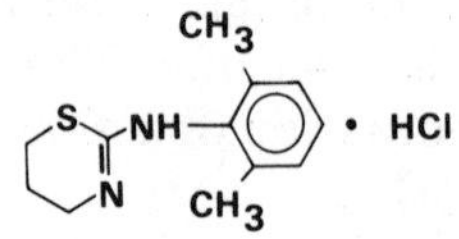

CH_3 S NH · HCl N CH_3

Fig. 14. Structure of xylazine.

Table 4
Time course of action of xylazine compared to dosage and pharmacokinetics

Species	Usual Dose, mg/kg	$t_{½}$,* min	V_D,* liter/kg	Cl_T,* $ml \cdot kg^{-1} \cdot min^{-1}$	Duration of Clinical Effects	Ref.
Swine	10.0				30–45 min	2
Dog	1.4	30	2.52	81	15–30 min	28
Sheep	1.0	23	2.74	83		
Horse	0.6	49	2.46	21	30–60 min	34
Cattle	0.2	36	1.94	42	24-h hyperglycemia	16
					36-h prostration	4

* See Table 1 for definitions. [Values for pharmacokinetic parameters are from Garcia-Villar et al. (17).]

relation between elimination rates and duration of action in dogs and horses, but the two seemed to be unrelated in the case of cattle. This species difference could be due to differences in receptors or responding systems or to an active metabolite. Further investigations are necessary to elucidate the pharmacology of xylazine in ruminants. The drug is an α_2-adrenoceptor agonist that has been shown to inhibit release of insulin from islet cells (34). The mechanism of its central nervous system effects has not been clarified. It seems to be the only drug particularly effective in obtunding severe pain in ruminant animals.

Clinical Management of Pain

One of the psychological curiosities of therapeutic decision making is the withholding of analgesic drugs, because the clinician is not absolutely certain that the animal is experiencing pain. Yet the same individual will administer antibiotics without documenting the presence of a bacterial infection. Pain and suffering constitute the only situation in which I believe that, if in doubt, one should go ahead and treat.

Suggestions of drugs for the alleviation of pain in domesticated animals are listed in Table 5. Aspirin would be given orally at the intervals indicated. Flunixin is generally administered intravenously to horses for colic and laminitis. Phenylbutazone may be given intravenously to cattle. Note, however, that neither phenylbutazone nor xylazine is approved by the Food and Drug Administration for use in food-producing animals. Even though salicylate has a short half-life in ruminants, it can be employed effectively in these animals by orally administering larger doses. Absorption from the rumen is rate limiting, and therapeutic concentrations are maintained in the plasma for 12 h after an oral dose of 100 mg/kg (19).

Table 5
Drugs recommended for control of pain in domesticated animals

Species	Mild to Moderate Somatic Pain		Severe Somatic or Visceral Pain	
	Drug	Dose, mg/kg	Drug	Dose, mg/kg
Dog	Aspirin	10, q 12 h	Morphine	0.25, q 6 h
Cat		10, q 48 h		0.1, q 6 h
Swine		10, q 4 h		0.2
Horse	Flunixin	1, iv	Methadone	0.25, q 6 h
Ruminant	Phenylbutazone	9, q 48 h	Xylazine	0.2
	Aspirin	100, q 12 h		

Morphine or methadone at the doses indicated are quite effective in controlling severe postsurgical pain or pain associated with trauma. The drugs are generally administered subcutaneously or intramuscularly as needed. The therapist should be familiar with the pharmacological actions of these drugs and their contraindications before administering them.

Conclusions

Considerable differences in the disposition of analgesic drugs are encountered among the domesticated animals. Knowledge of these and appropriate modifications of dosage regimens allow the veterinarian to provide effective relief of pain in these animals with drugs that are readily available. This enables the veterinarian to fulfill his ethical obligations to these patients.

REFERENCES

1. Alexander, F., and R. A. Collett. Pethidine in the horse. *Res. Vet. Sci.* 17: 136–137, 1974.
2. Benson, G. J., and J. C. Thurmon. Anesthesia of swine under field conditions. *J. Am. Vet. Med. Assoc.* 174: 594–596, 1979.
3. Boulos, B. M., W. L. Jenkins, and L. E. Davis. Pharmacokinetics of certain drugs in the domesticated goat. *Am. J. Vet. Res.* 33: 943–952, 1972.
4. Clarke, K. W., and L. W. Hall. Xylazine. A new sedative for horses and cattle. *Vet. Rec.* 85: 512–517, 1969.
5. Davis, L. E. A perspective on drug therapy in veterinary medicine. In: *Proc. Equine Pharmacol. Symp., 2nd,* edited by J. D. Powers and T. E. Powers. Golden, CO: Am. Assoc. Equine Practitioners, 1978, p. 9–16.
6. Davis, L.E. The challenge of veterinary pharmacology. *Trends Pharmacol. Sci.* 1: 295–299, 1980.
7. Davis, L. E., C. A. Davis, and J. D. Baggot. Comparative pharmacokinetics in domesticated animals. In: *Research Animals in Medicine,* edited by L. H. Harmison. Bethesda, MD: NIH, 1973, 72–333, p. 715–733.
8. Davis, L. E., and E. J. Donnelly. Analgesic drugs in the cat. *J. Am. Vet. Med. Assoc.* 153: 1161–1167, 1968.

9. Davis, L. E., and W. L. Jenkins. Some considerations regarding drug therapy in ruminant animals. *Bovine Pract.* 9: 57–60, 1974.
10. Davis, L. E., C. A. Neff-Davis, and J. R. Wilcke. Monitoring drug concentrations in animal patients. *J. Am. Vet. Med. Assoc.* 176: 1156–1158, 1980.
11. Davis, L. E., and B. L. Sturm. Plasma levels and effects of pentazocine in domesticated animals. *Am. J. Vet. Res.* 31: 1631–1635, 1970.
12. Davis, L. E., and B. A. Westfall. Species differences in the biotransformation and excretion of salicylate. *Am. J. Vet. Res.* 33: 1253–1262, 1972.
13. Dayton, P. G., Z. H. Israili, and J. M. Perel. Influence of binding on drug metabolism and distribution. *Ann. NY Acad. Sci.* 226: 172–194, 1973.
14. DeBacker, P., R. Braeckman, F. Belpaire, and M. Debackere. Bioavailability and pharmacokinetics of phenylbutazone in the cow. *J. Vet. Pharmacol. Ther.* 3: 29–33, 1980.
15. Eberhardson, B., G. Olsson, L. Appelgren, and S. Jacobsson. Pharmacokinetic studies of phenylbutazone in cattle. *J. Vet. Pharmacol. Ther.* 2: 31–37, 1979.
16. Eichner, R. D., R. L. Prior, and W. G. Kuasnicka. Xylazine-induced hyperglycemia in beef cattle. *Am. J. Vet. Res.* 40: 127–129, 1979.
17. Garcia-Villar, R., P. L. Toutain, M. Alvinerie, and Y. Ruckebusch. The pharmacokinetics of xylazine hydrochloride: an interspecific study. *J. Vet. Pharmacol. Ther.* 4: 87–92, 1981.
18. Garrett, E. R., and T. Gurkan. Pharmacokinetics of morphine and its surrogates. IV. Pharmacokinetics of heroin and its derived metabolites in dogs. *J. Pharm. Sci.* 69: 1116–1134, 1980.
18a. Getty, R. *Sisson and Grossman's The Anatomy of the Domestic Animals* (5th ed.). Philadelphia, PA: Saunders, 1975, vol. I.
19. Gingerich, D. A., J. D. Baggot, and R. A. Yeary. Pharmacokinetics and dosage of aspirin in cattle. *J. Am. Vet. Med. Assoc.* 167: 945–948, 1975.
20. Hazum, E., J. J. Sabatka, K. Chang, D. A. Brent, J. W. A. Findlay, and P. Cuatrecasas. Morphine in cow and human milk: could dietary morphine constitute a ligand for specific morphine (μ) receptors? *Science* 213: 1010–1012, 1981.
21. Hucker, H. B. Species differences in drug metabolism. *Annu. Rev. Pharmacol.* 10: 99–118, 1970.
22. Hug, C. C., Jr., M. R. Murphy, E. P. Rigel, and W. A. Olson. Pharmacokinetics of morphine injected intravenously into the anesthetized dog. *Anesthesiology* 54: 38–47, 1981.
23. Hvidberg, E. F., and F. Rasmussen. Pharmacokinetics of phenylbutazone and oxyphenbutazone in the pig. *Can. J. Comp. Med.* 39: 80–88, 1975.
24. Jenkins, W. L. *Drug Distribution Across the Ruminal Epithelium.* Columbia: Univ. of Missouri, 1969. PhD thesis.
25. Jenkins, W. L., and L. E. Davis. Transfer of drugs across the ruminal wall of goats. *Am. J. Vet. Res.* 36: 1771–1776, 1975.
26. Loan, W. B., and J. D. Morrison. Strong analgesics: pharmacological and therapeutic aspects. *Drugs* 5: 108, 1973.
27. Marriner, S., and J. A. Bogan. The influence of the rumen on the absorption of drugs: studies using neclofenamic acid administered by various routes to sheep and cattle. *J. Vet. Pharmacol. Ther.* 2: 109–115, 1979.
28. Newkirk, H. L., and D. G. Miles. Xylazine as a sedative analgesic for dogs and cats. *Mod. Vet. Pract.* 55: 677–680, 1974.
29. Palminteri, A. Oxymorphone, an effective analgesic in dogs and cats. *J. Am. Vet. Med. Assoc.* 143: 160–163, 1963.
30. Piperno, E., D. J. Ellis, S. M. Getty, and T. M. Brody. Plasma and urine levels of phenylbutazone in the horse. *J. Am. Vet. Med. Assoc.* 153: 195–198, 1968.

31. Richez, P., A. Regnier, and Y. Ruckebush. Influence of food intake on the absorption of lysine-acetylsalicylate in dogs. *J. Vet. Pharmacol. Ther.* 3: 121–124, 1980.
32. Spector, S. Disposition of drugs in man by radioimmunoassay. *Pharmacol. Rev.* 34: 73–75, 1982.
33. Stevens, C. E. Comparative physiology of the digestive system. In: *Dukes' Physiology of Domestic Animals* (9th ed.), edited by M. J. Swenson. Ithaca, NY: Comstock, 1977, p. 222.
34. Thurmon, J. C., C. A. Neff-Davis, L. E. Davis, R. A. Stoker, G. J. Benson, and T. F. Lock. Xylazine hydrochloride induces hyperglycemia and hypoinsulinemia in thoroughbred horses. *J. Vet. Pharmacol. Ther.* 5: 241–245, 1982.
35. Vernimb, G. D., and P. W. Hennessey. Clinical studies on flunixin meglumine in the treatment of equine colic. *J. Equine Med. Surg.* 1: 111–116, 1977.

ELEVEN

Evaluation of Analgesic Drugs in Horses

William V. Lumb

Department of Clinical Sciences, College of Veterinary Medicine and Biomedical Sciences, Colorado State University, Fort Collins, Colorado

Ney L. Pippi

Escola de Veterinaria, Universidade Federal de Santa Maria, 97.100 Santa Maria, Rio Grande do Sul, Brazil

Marissak Kalpravidh

Faculty of Veterinary Sciences, Department of Surgery, Chulalongkorn University, Bangkok, Thailand

Of the domesticated species of animals, horses are probably most refractory to effects of analgesic drugs. At the same time, a great need exists for effective agents that produce a sedative and analgesic effect in this species. A striking example of this occurred in 1975 when the magnificent filly, Ruffian, broke her leg in a match race at New York's Belmont Park. She was treated by a team of four veterinarians and an orthopedic surgeon. During recovery from anesthesia she struggled desperately and destroyed the fixation device that had been applied. A decision for euthanasia was made 8 h after the race.

Horses with colic are restless, tend to kick and bite at handlers, and are generally hazardous for veterinarians to examine for diagnosis of the colic cause. These animals need to be relieved of their pain and sedated for an adequate examination. In these horses and those awakening from anesthesia, alleviation of pain can minimize the risk of complications, facilitate treatment, and decrease the chances for injury to the animal and to personnel.

In past years, efforts were made to assess the effects of analgesics in horses by subjective means. Lowe (40) induced visceral pain in ponies by inflating a rubber balloon in the cecum. The clinical signs produced simulated those of flatulent colic, and the effects of analgesic drugs were assessed visually after their administration. Tobin and colleagues (10, 57) developed a method for determining behavioral responses to narcotic analgesics. Their system consisted of putting the animal in a closed box stall after narcotic administration. The stall contained a rack with hay but was otherwise vacant. A 1-ft-square glass window was installed for observation of the horse. After intravenous administration of the drug, an observer recorded the eating behavior and the number of steps taken by the horse for 2 min. The need for more objective means of assessing the efficacy of analgesic drugs in horses led to development of other models of pain.

Models to Test Analgesics in Horses

Pippi model

In 1979, Pippi et al. (51) devised a model for evaluating superficial, deep, and visceral pain in ponies. This was the first experimental model in which analgesic drugs were objectively tested in equine species.

To produce superficial pain, a modification of the Hardy-Wolff-Goodell method (24), commonly used in humans, was developed. A 1,000-W quartz lamp was directed to an area of skin just above the coronary band of the forelimb. The lamp was focused on the skin, which had been previously painted with lamp black, and an accelerometer on a Velcro-covered band was placed around the forelimb above the knee. When the hot light was directed on the skin, the animal would lift the foot, activating the accelerometer, which in turn shut off a timing device. Reaction time to the stimulus was read from the timer.

Radiant stimuli, either of variable intensity-fixed time or fixed intensity-variable time, tend to damage tissues. This limits the number of valid stimulations at a given site (35). For this reason, individual tests were conducted at slightly different anatomical locations. The skin was painted with lamp black to ensure total absorption of radiation regardless of pigmentation.

To produce deep pain, a device was developed consisting of an implantable heating element 2 cm long and 0.5 cm in diameter and connected to wires encased in a Silastic sheath. This device was surgically implanted on the midlateral periosteal surface of the radius. During tests, the cable was connected to a power supply. The power delivered to the device was monitored by a volt and current meter.

Power delivered to the device was maintained at 150 W, by a current of 12.5 A at 12 V. The accelerometer was again attached to the forelimb above the carpus to measure reaction time.

Deep pain is not well localized (19), which is probably why reaction to the heating device on the periosteum of the radius was slow. This type of pain stimulus proved to be the least reliable of the three utilized (Table 1). In addition, difficulties were experienced with long-term implants. These tended to become infiltrated with tissue fluids, short circuiting the attached wires and making the tests impossible.

For visceral pain a modification was made of Lowe's (40) balloon-induced colic model. The balloon was used, but in addition an accelerometer mounted on a girth around the abdomen was attached to detect the colic-type movements produced by the animal. Reaction of the accelerometer to the horse's movement was coupled electronically to a recorder, and a 5-mm deflection of the channel marker was considered to be a threshold movement indicating colic pain (Fig. 1).

Visceral pain is elicited by distention of the bowel accompanied by a significant change in tension (38, 39). In our pain model, pressure is increased within the cecum, thus eliciting visceral pain and typical colic symptoms in 1–3 min. Pressure had to be constant (100 mmHg) in all animals for good results, in contrast to the technique of Lowe (personal communication) in which pressure always fell to 10–50 mmHg. The pain threshold (recorder line increase of 5 mm) was fixed after preliminary testing with several animals. This method required a vigorous body movement and was effective in eliminating extraneous movements commonly occurring with horses standing in stocks.

Table 1

Mean reaction times in seconds of ponies subjected to three types of pain

Pony Number	Superficial Pain	Deep Pain	Visceral Pain
1	8.2 ± 6.0	189.48 ± 85.81	161.5 ± 22.74
2	6.11 ± 1.66	24.30 ± 6.63	126.12 ± 10.39
3	5.96 ± 0.63	111.13 ± 31.54	145.5 ± 39.96
4	4.17 ± 0.35	32.26 ± 17.57	66.87 ± 13.06
5	5.30 ± 0.96	49.42 ± 34.50	148.12 ± 26.63
6	6.38 ± 0.72	36.27 ± 24.02	78.37 ± 27.21
7	6.18 ± 0.65	113.11 ± 33.88	202.50 ± 18.49
8	13.78 ± 9.17	91.07 ± 62.67	113.75 ± 28.31
9	29.29 ± 11.80	181.23 ± 99.51	84.25 ± 19.89
10	5.23 ± 0.49	73.45 ± 58.45	161.50 ± 18.66
Mean	9.06 ± 5.14	90.17 ± 53.73	128.84 ± 23.94

Values are means ± SD for 8 measurements (every 30 min) for 4 h. [Data from Pippi, Lumb, et al. (51).]

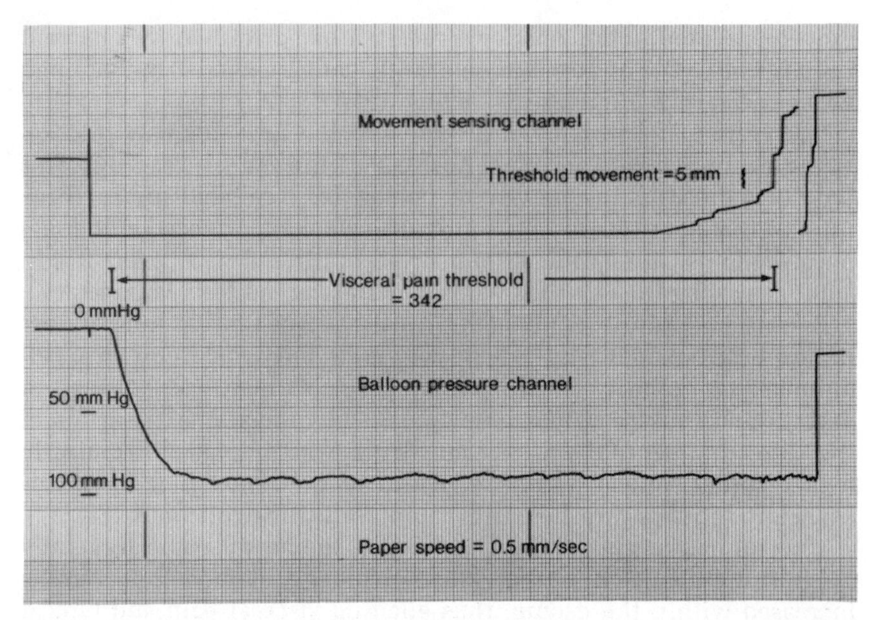

Fig. 1. Recording of visceral pain test. Upper channel sensed movement of animal, detected by girth-mounted accelerometer. A 5-mm deflection was considered threshold movement. Pain threshold was determined in seconds from initiation of balloon inflation. Lower channel indicates pressure within balloon.

Variability in pain thresholds is a common problem with models designed for other species, including humans (6, 7, 35, 43, 48). With our model the three kinds of pain had clearly defined end points as required by Beecher (7). It also fulfilled the requirements for producing painful stimuli outlined by Goetzl et al. (21), Chapman and Jones (6), Miller (47), Hardy et al. (23), and Lahoda et al. (35), except that it is not applicable to human subjects.

Brunson model

More recently, Brunson et al. (4) developed a method of dental dolorimetry for pain-threshold studies in the horse. A tooth-pulp technique was chosen because innervation of teeth is presumably of Aδ- and C-fiber types. A monopolar electrode was implanted in the dentine layer of the horse's canine tooth and secured with an insulating dental acrylic. Teflon-coated wire from the electrode was passed under the gingiva and brought through the skin on the cheek, or, in the case of the lower canine tooth, the wire exited through the skin on the lateral side of the mandible. An electric current was applied to the tooth pulp from a constant current stimulator. The current was measured with an oscilloscope, enabling precise quantification of the stimulus, which was 2.0 ms in duration and was repeated at 20-s intervals three times at each current level. Upward movement of the horse's head indicated a positive response to the stimulus. Three consecutive positive responses were required to indicate the pain threshold. The animal was tested every 10 min after injection of a drug. These investigators indicated that consistent and reproducible pain-threshold measurements could thus be determined in the horse. The threshold did not vary with repetitive stimulation, time of day, or from one day to the next. Conditioning of the subject was not apparent, and implants lasted as long as 6 mo. Data from this model have not been published.

Drugs Tested With Pippi Model

Opiate analgesic drugs act on specific opiate receptors in the brain and spinal cord, raising the pain threshold. Unfortunately these effects are inconsistent in horses; the animals show varying degrees of analgesia and sedation from these agents, or become restless, ataxic, and/or excited. We have investigated the effects of nine drugs with our equine pain model. Some had previously been used clinically in horses, others had not.

Morphine is a prototype for the opiate analgesic drugs. It has been used in the horse to relieve the acute pain of spasmodic colic (2), but it is not widely accepted because the response is unreliable. Some horses are relieved, whereas many show excitement that is undesirable and dangerous. It has been recommended that no more than 60 (22), 120 (2), or 200 mg (55) be given to a horse. Morphine given intravenously induced eating behavior at a dose of 1.2 mg/kg and caused increases in locomotor activity (10). The onset of incoordination was 1 h and lasted up to 7 h.

Fentanyl citrate, a synthetic phenylpiperidine-derivative analgesic, has been used alone or in combination with a tranquilizer (droperidol) for neuroleptanalgesia (11, 20, 29, 33, 37, 44). A dose of 0.2 mg/kg of body weight intravenously produces recumbency in ponies but not surgical anesthesia (D. A. Gingerich, personal communication). A dose of 0.1 mg/kg was suggested for trial by the manufacturer.

Levorphanol tartrate, a morphinan derivative, is a synthetic analgesic with properties and actions similar to morphine (5, 16). It is 3–5 times more potent than morphine (3, 17, 25, 31, 36). It has not been approved for clinical use in horses. The average dose in humans is 2–3 mg subcutaneously with maximal analgesia occurring at 60–90 min postinjection and a duration of 5.5–7 h (31). It provided satisfactory analgesia for postoperative patients without or with minimal respiratory depression (3, 17, 18, 30).

Meperidine hydrochloride, a synthetic phenylpiperidine-derivative analgesic, has been used as a restraint adjuvant and preanesthetic in several species. Lumb and Jones (41) recommended 1.1–4.4 mg/kg intravenously for preanesthesia in horses. Lowe (personal communication) used meperidine (2.2 mg/kg) intramuscularly with other drugs in a blind study to relieve induced colic in ponies.

Methadone hydrochloride, a synthetic diphenylheptane-derivative narcotic analgesic, has been used in combination with phenothiazine tranquilizers in horses (41). Schauffler (54) utilized methadone plus acetylpromazine as preanesthetic in doses of 0.11 mg (methadone)/kg for horses.

Oxymorphone hydrochloride, a semisynthetic phenanthrene-derivative analgesic, has been used subcutaneously and intramuscularly in small animals (49). Palminteri (50) employed it in dogs (0.66 mg/kg) and, in combination with a tranquilizer, in cats as a preanesthetic agent.

Pentazocine lactate, structurally related to phenazocine, is a synthetic benzomorphan analgesic with opiate agonist and antagonist properties. Lowe (40) used from 0.55 to 1.1 mg/kg intravenous in ponies for short duration analgesia and 2.2, 4.4, or 6.6 mg/kg intramuscular for prolonged analgesia. Davis and Sturm (12) used 6.6 mg/kg in ponies.

Butorphanol tartrate, a synthetic morphinan derivative chemically related to naloxone, is a centrally acting analgesic with both opiate agonist and antagonist properties (52). In comparative studies of agonist activity with intramuscular doses in humans with acute pain, butorphanol was ~7 and ~20 times as potent as morphine sulfate (13, 15, 56) and pentazocine lactate (14), respectively. In rodents, butorphanol was 4 and 30 times more potent than morphine and pentazocine, respectively. As a narcotic antagonist in reversing the effect of oxymorphone or morphine in rodents, subcutaneous butorphanol was 30 times more potent than pentazocine, about equipotent with nalorphine, and ~50 times less potent than naloxone (52). Doses of 0.022 mg/kg have been tried by Reed and Bayly (53) in horses with an observed duration of effects of 2 h.

Xylazine hydrochloride is 2-(2,6-dimethylphenylamino) 4*H*-5,6-dihydro-1,3-thiazine hydrochloride. It is a nonopiate, centrally acting sedative analgesic with muscle-relaxant properties. Its depressant action is mediated by α_2-adrenergic receptors (28). The drug can be given intravenously to horses at 1.1 mg/kg and intramuscularly at 2.2 mg/kg (34). Clarke and Hall (9) used 2.3 mg/kg xylazine in horses. McCashin and Gabel (46) recommended 2 mg/kg administered intramuscularly in horses for sedative and preanesthetic effects. In evaluation of analgesia in 158 horses, results were excellent in 75, good in 54, fair in 17, and poor in 12 (26). Klein and Baetjer (32) tested a combination of xylazine (mean dose 1.2 mg/kg) and morphine (mean dose 0.75 mg/kg) given intravenously in 16 horses. Restraint was considered superior to that achieved from xylazine alone. Side effects attributed to morphine included sweating, tremors of the muzzle and tongue, and cutaneous wheals. Mackenzie (42) studied combinations of etorphine and various tranquilizers and believed that etorphine-xylazine was the best combination. Lowe (personal communication) compared the analgesic actions of xylazine, pentazocine, meperidine, and dipyrone with the cecal-balloon technique in five ponies. He concluded that xylazine at 2.2 mg/kg was the only effective drug for producing analgesia.

Flunixin meglumine is an antiprostaglandin of the fenamic acid group (8). It is a peripherally acting nonnarcotic, nonsteroidal agent with analgesic, anti-inflammatory, and antipyretic activity in the horse at a low concentration (27). A single parenteral or oral dose of 1.1 mg/kg of body weight produced maximal analgesic and anti-inflammatory actions in a standardized equine lameness model. An intravenous dose decreased lameness and swelling by 66 and 34%, respectively, at 30 h postinjection. Intramuscular flunixin at 0.55 and 1.1 mg/kg produced comparable efficacy to similar intravenous doses. Oral flunixin improved lameness and swelling by 71 and 31%, respectively. Onset and duration of activity were not affected by route of administration. Onset occurred within 2 h after administration; peak response was seen at 12 h, and activity persisted for 30 h.

Flunixin has a wide margin of safety in horses (27). Anorexia and postinjection trembling were the only side effects reported after parenteral or oral administration of more than 700 mg of flunixin.

Clinical studies demonstrated that 1.1 mg/kg of parenteral or oral flunixin was efficacious for alleviation of pain and inflammation associated with musculoskeletal disorders in 74% of 262 horses (27), and for colic in 68% of 118 cases (58). Vernimb and Hennessey (58) reported that improvement of colic symptoms was noted in 38% of the horses within 15 min and in 89% by 60 min. Duration of action in colic horses averaged 6–8 h in horses subject to recurrence. Best results were obtained in animals affected by flatulent or spastic colic. Oral or parenteral flunixin in horses was four times as potent as phenylbutazone (27).

Experiment I

Materials and methods

Six ponies of both sexes and breeds weighing between 111 and 263 kg and from 2 to 7 yr of age were used. The Pippi model was used to

Table 2

Design for experiment I

Day	Nature of Work
1	Implant deep-pain device on radius Create cecal fistula
21	Determine control values for pain Superficial Deep Visceral
22–27	Administer drugs in Latin square: one drug each day; test each drug for 4 h at 30-min intervals
28	Determine control values for pain Superficial Deep Visceral

test their reaction to superficial, deep, and visceral pain (Tables 1 and 2; 51).

Six drugs were tested in a Latin-square design. With the exception of fentanyl and oxymorphone, which were determined by preliminary testing, clinically recommended doses were used: fentanyl citrate, 0.22 mg/kg; meperidine hydrochloride, 4.4 mg/kg; methadone hydrochloride, 0.22 mg/kg; oxymorphone hydrochloride, 0.033 mg/kg; pentazocine lactate, 2.2 mg/kg; xylazine hydrochloride, 2.2 mg/kg. Control and test values for superficial, deep, and visceral pain thresholds were measured in random sequence every 30 min for 4 h; the tests for three types of pain were in random sequence.

Results

Means of the mean values of all six ponies for the three kinds of pain are tabulated at 2- and 4-h intervals (Table 3). Rank values are shown in Table 4. Significant differences between drugs tested for superficial pain and control values were not detected by variance analysis of the means at 2- and 4-h intervals. Xylazine was significantly better ($P < 0.05$) than other drugs tested for relief of deep pain at 2 and 4 h (4 and 8 measurements). Significant differences between drug and control values for relief of visceral pain were not detected at 2- or 4-h intervals.

Experiment II

Materials and methods

In a group of four ponies xylazine was tested again. In addition, a combination of this drug with each of the three next most effective

Table 3
Mean of mean control and drug values in seconds for six ponies

	Superficial Pain		Deep Pain		Visceral Pain	
	2 h	4 h	2 h	4 h	2 h	4 h
Control, day 1	6.20	6.02	70.56	73.82	123.04	121.08
Fentanyl	11.20	9.25	118.51	92.19	169.60	150.79
Meperidine	7.23	6.27	220.72	165.11	170.45	155.25
Methadone	5.72	5.25	108.13	88.67	163.54	138.68
Oxymorphone	5.84	5.55	176.21	131.44	189.00	160.22
Pentazocine	6.80	6.28	141.89	114.21	164.62	152.83
Xylazine	26.35	17.08	430.61	312.23	200.16	171.26
Control, day 8	4.39	4.34	41.62	41.19	114.33	106.43

Figures for 2-h intervals represent mean of 4 values obtained at 30-min intervals. Figures for 4-h intervals represent mean of 8 values obtained at 30-min intervals. [Adapted from Pippi and Lumb (50a).]

Table 4
Performance ranking of drugs in six ponies

Treatment	2-h Intervals (4 Measurements)							4-h Intervals (8 Measurements)						
	Pony number						Total of rank values	Pony number						Total of rank values
	1	2	3	4	5	6		1	2	3	4	5	6	
Superficial pain														
Fentanyl	6	5	1	3	4	5	24	6	5	1	3	4	5	24
Meperidine	2	2	5	5	3	4	21	2	2	5	4	3	4	20
Methadone	1	4	4	4	1	1	15	1	4	4	6	1	1	17
Oxymorphone	3	1	3	1	2	3	13	3	1	2	2	2	2	12
Pentazocine	4	3	2	2	5	2	18	4	3	3	1	5	3	19
Xylazine	5	6	6	6	6	6	35	5	6	6	5	6	6	34
Deep pain														
Fentanyl	2	2	2	3	5	3	17	2	3	2	3	5	3	18
Meperidine	5	4	5	5	4	4	27	5	4	5	5	4	4	27
Methadone	3	3	4	3	3	1	15	3	2	4	2	1	1	13
Oxymorphone	4	5	3	6	1	2	21	4	5	3	6	3	2	23
Pentazocine	1	1	1	1	3	6	13	1	1	1	1	2	6	12
Xylazine	6	6	6	4	6	5	33	6	6	6	4	6	5	33
Visceral pain														
Fentanyl	4	3	2	3	1	6	19	1	3	2	3	2	6	17
Meperidine	5	2	3	5	4	2	21	6	2	3	6	5	3	25
Methadone	6	1	5	1	2	3	18	5	1	4	1	1	2	14
Oxymorphone	1	5	6	6	3	1	22	4	5	6	4	3	1	23
Pentazocine	2	4	1	2	5	4	18	2	4	1	5	6	4	22
Xylazine	3	6	4	4	6	5	28	3	6	5	2	4	5	25

Best drug ranked 6, with values decreasing to poorest ranked 1. [From Pippi and Lumb (50a).]

drugs was tested, utilizing a four-by-four Latin square. The Pippi model was used again; control values were obtained on the 21st day after surgery with testing done on the following 4 days. Final control values were obtained on the 26th postoperative day. The 30-min interval for testing used in experiment I was again employed.

Half doses of the drugs used in experiment I were used in combination with xylazine in experiment II with the exception of fentanyl. It was used at one-fourth of the original dosage, because a half dose of fentanyl combined with a half dose of xylazine produced excitation in two ponies. Drugs and the intramuscular doses used were: xylazine hydrochloride, 2.2 mg/kg; xylazine hydrochloride, 1.1 mg/kg, plus fentanyl citrate, 0.055 mg/kg; xylazine hydrochloride, 1.1 mg/kg, plus meperidine hydrochloride, 2.2 mg/kg; xylazine hydrochloride, 1.1 mg/kg, plus oxymorphone hydrochloride, 0.0165 mg/kg.

Results

Rank values for experiment II are shown in Table 5. A significant difference in pain threshold for superficial pain was not seen at either time interval (2 or 4 h). In the two time intervals, no drug differed

Table 5
Performance ranking of drugs in four ponies

Treatment	2-h Intervals (4 Measurements)					4-h Intervals (8 Measurements)				
	Pony number				Total of rank values	Pony number				Total of rank values
	1	2	3	4		1	2	3	4	
Superficial pain										
Xylazine	1	1	1	2	5	2	1	2	2	7
Xylazine + fentanyl	2	4	4	3	13	1	4	4	3	12
Xylazine + meperidine	4	3	2	4	13	4	3	1	4	12
Xylazine + oxymorphone	3	2	3	1	9	3	2	3	1	9
Deep pain										
Xylazine	3	3	3	4	13	3	3	3	4	13
Xylazine + fentanyl	1	1	4	1	7	1	2	4	1	8
Xylazine + meperidine	4	4	2	3	13	4	4	2	3	13
Xylazine + oxymorphone	2	2	1	2	7	2	1	1	2	6
Visceral pain										
Xylazine	1	3	4	4	12	2	3	3	4	12
Xylazine + fentanyl	3	4	3	3	13	3	4	4	3	14
Xylazine + meperidine	4	1	1	2	8	4	1	1	2	8
Xylazine + oxymorphone	2	2	2	1	7	1	2	2	1	6

Best drug ranked 4, with values decreasing to poorest ranked 1. [From Pippi and Lumb (50a).]

significantly from the other for relieving deep pain. The combination of xylazine and fentanyl was significantly better ($P < 0.10$) for obtunding visceral pain than xylazine alone or the other drug combinations at the 2- and 4-h intervals.

There was considerable variability from pony to pony for the measured values of the same drugs. In terms of subjective evaluation, variation was also observed. Some ponies seemed well sedated and others were not.

Experiment III

Materials and methods

Eight mixed-breed ponies, one male and seven females, weighing 111–163 kg and ranging from 4 to 18 yr of age were used. Three weeks before testing, surgery was performed to implant a cecal cannula and to create a subcutaneous carotid loop. Three days before testing, a 10-cm polyethylene catheter was inserted into this loop for measuring arterial blood pressures and obtaining arterial blood samples for blood gas evaluation. A cannula was also inserted in the jugular vein with its tip near the heart for measuring central venous pressure. The Pippi model was used to induce superficial and visceral pain (51).

A six-channel recorder was used for registering electrocardiograms, blood pressures, balloon pressures, and movements of the animal. Pressure transducers were used to measure blood and balloon pressures. Mean pressure was obtained by adding one-third of pulse pressure (systolic pressure – diastolic pressure) to diastolic pressure. The respiratory rate was grossly observed, while body temperature was read from a rectal thermometer. An acid-base analyzer was used for quantitating arterial blood pH (pH_a), oxygen tension (Pa_{O_2}), carbon dioxide tension (Pa_{CO_2}), and bicarbonate (HCO_3^-).

The drugs and doses given intramuscularly were: butorphanol tartrate, 0.22 mg/kg; flunixin meglumine, 2.2 mg/kg; levorphanol tartrate, 0.033 mg/kg; morphine sulfate, 0.66 mg/kg; xylazine hydrochloride, 2.2 mg/kg. A different sequence of drug administration was scheduled for each pony from the 22nd to 26th postoperative day, and a single drug was given each day. All parameters were sequentially measured before drug injection (daily control value), and after injection at 30, 60, 120, 180, and 240 min.

Student's *t* test was used to analyze the results. The individual relative changes of pain threshold [postinjection threshold – preinjection threshold (daily control)/preinjection threshold] after drug injections were used to evaluate and compare the analgesic effects. Significant difference ($P < 0.10$) in postinjection pain threshold from preinjection threshold indicated that the drug had an analgesic effect. The individual relative change of each cardiopulmonary parameter after drug injection was used to justify the significant effects of the drug.

Results

Analgesic effect. No significant analgesic effect was observed for superficial pain after flunixin or levorphanol was administered (Fig. 2). Butorphanol produced detectable analgesia at 60 and 180 min. After morphine administration, significant effects were seen at 30, 60, and 120 min. Significant analgesic effects produced by xylazine were observed at 30, 60, 120, and 180 min postinjection.

Morphine was better than butorphanol at 60 and 120 min and better than flunixin at 30, 60, and 120 min. It was superior to levorphanol at 30, 60, 120, and 180 min. No significant difference was observed between morphine and xylazine in relieving superficial pain. Superior effects of xylazine over butorphanol were detected at 30, 60, and 120 min. Xylazine was better than flunixin and levorphanol at 30, 60, 120, and 180 min. Significant differences between butorphanol, flunixin, and levorphanol were not observed at any interval.

Levorphanol, xylazine, and butorphanol significantly increased

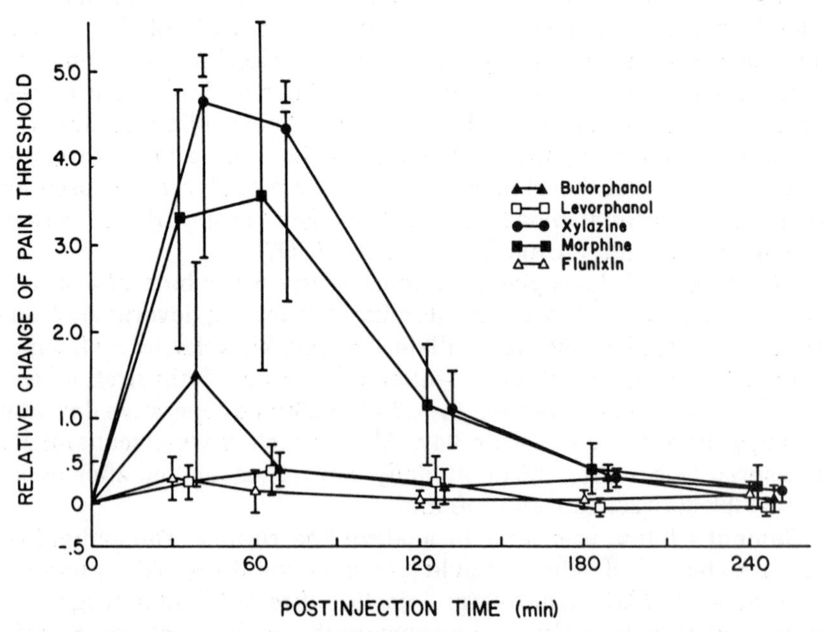

Fig. 2. Relative changes in threshold of superficial pain after drug administration. Each point represents mean ±SEM.

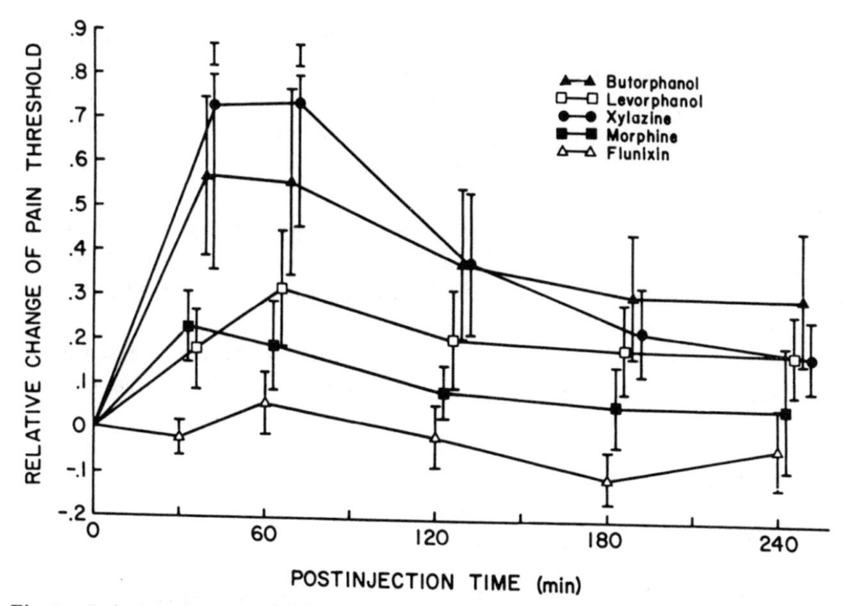

Fig. 3. Relative changes in threshold of visceral pain after drug administration. Each point represents mean ±SEM.

thresholds for visceral pain at 30, 60, 120, 180, and 240 min, whereas no significant change was observed after flunixin administration (Fig. 3). Morphine produced effects at 30, 60, and 120 min postinjection.

The effect of butorphanol did not differ from that of xylazine. Butorphanol was better than flunixin at 30, 60, 120, 180, and 240 min and levorphanol at 30 and 60 min. Butorphanol was superior to morphine at 30, 60, 120, and 180 min. Levorphanol was better than flunixin at 30, 60, 120, 180, and 240 min. There was no significant difference between levorphanol and morphine. The superior effects of morphine over flunixin were seen only at 30 and 180 min. Xylazine was significantly better than flunixin at 30, 60, 120, 180, and 240 min. It was also better than morphine at 30, 60, 120, and 180 min and levorphanol at 30 and 60 min.

Effects on cardiopulmonary system. Complete data were evaluated from six ponies. Values for heart rate, blood pressures, respiration rate, body temperature, and blood gas variables were averaged from 30 preinjection measurements in six ponies. Means and standard deviations of these base-line values were: heart rate, 57 ± 9 beats/min; central venous pressure, −9 ± 6 mmHg; systolic pressure, 130 ± 14 mmHg; diastolic pressure 73 ± 10 mmHg; and mean pressure, 92 ± 10 mmHg. Mean respiratory rate and body temperature were 19 ± 12 breaths/min and 38.5 ± 0.5°C. Means and standard deviations of blood gas variables were: Pa_{O_2}, 79 ± 5.5 mmHg; Pa_{CO_2}, 34 ± 3.5 mmHg; pH_a, 7.43 + 0.03; and HCO_3^-), 22 ± 2.5 meq/liter.

Cardiac arrhythmias and changes outside the normal limits of body temperature were not observed after injecting any of the drugs. Cardiopulmonary changes after flunixin or levorphanol administration were insignificant.

Effects of butorphanol on heart rate and blood pressures are shown in Figure 4. Butorphanol significantly increased heart rate at 30 and 60 min postinjection. Changes in respiratory rate and blood gas variables were still within their normal ranges.

After morphine administration, heart rate and blood pressures increased (Fig. 5). The slight increase in central venous pressure was not

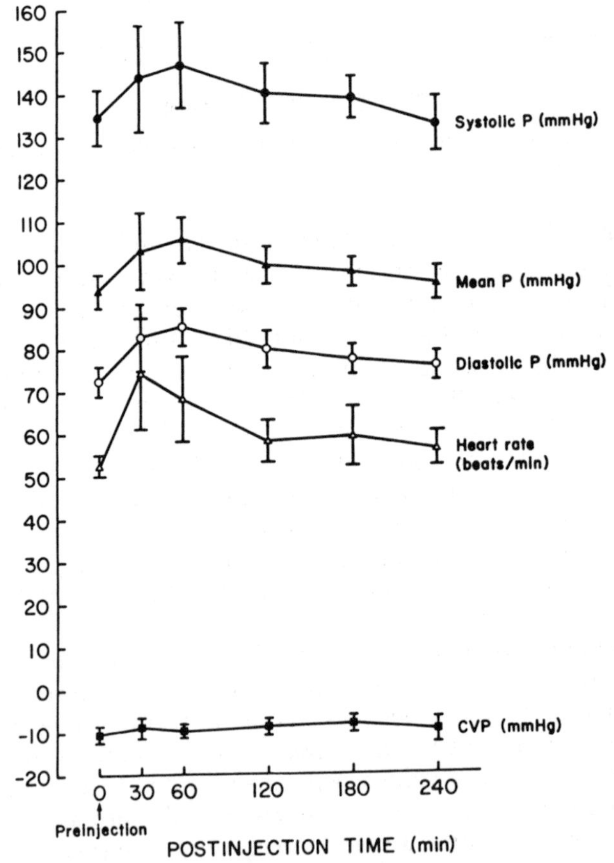

Fig. 4. Changes in heart rate and blood pressure after butorphanol administration (0.22 mg/kg). Each point represents mean ±SEM.

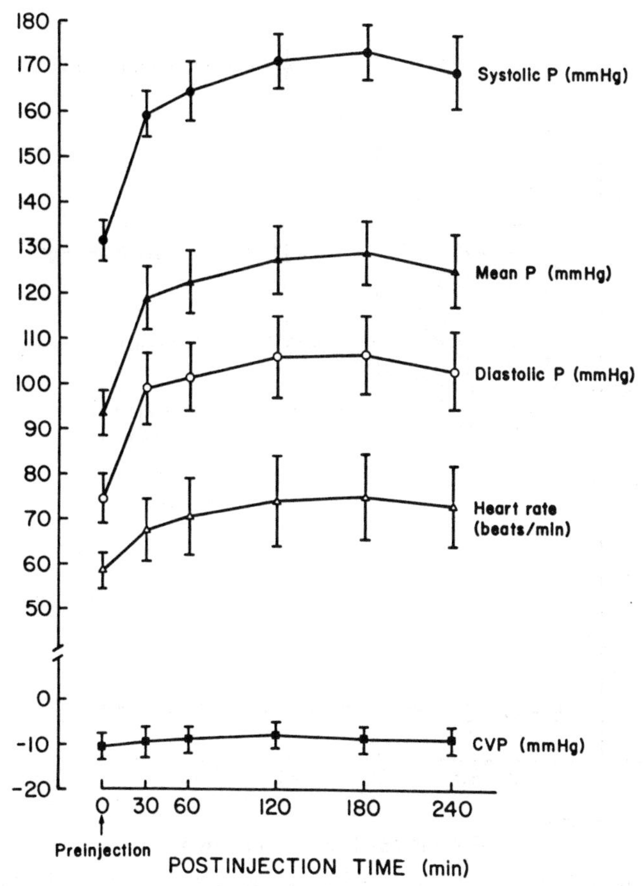

Fig. 5. Changes in heart rate and blood pressures after morphine administration (0.66 mg/kg). Each point represents mean ±SEM.

significant, but the effects of morphine on heart rate were significant at 120, 180, and 240 min. Morphine increased systolic, diastolic, and mean pressures at 30, 60, 120, 180, and 240 min. Respiratory rates were increased at 60, 120, 180, and 240 min. Changes in blood gas variables remained within normal limits. Effects of xylazine on heart rate and blood pressures are shown in Figure 6. Decreases in heart rate observed at every interval were still within the normal range. Central venous pressure was not significantly affected by xylazine; whereas systolic, diastolic, and mean blood pressures were decreased at 30 and 60 min. Changes in respiratory rate and blood gas variables remained within normal ranges.

Behavioral effects. All eight ponies in experiment III were observed

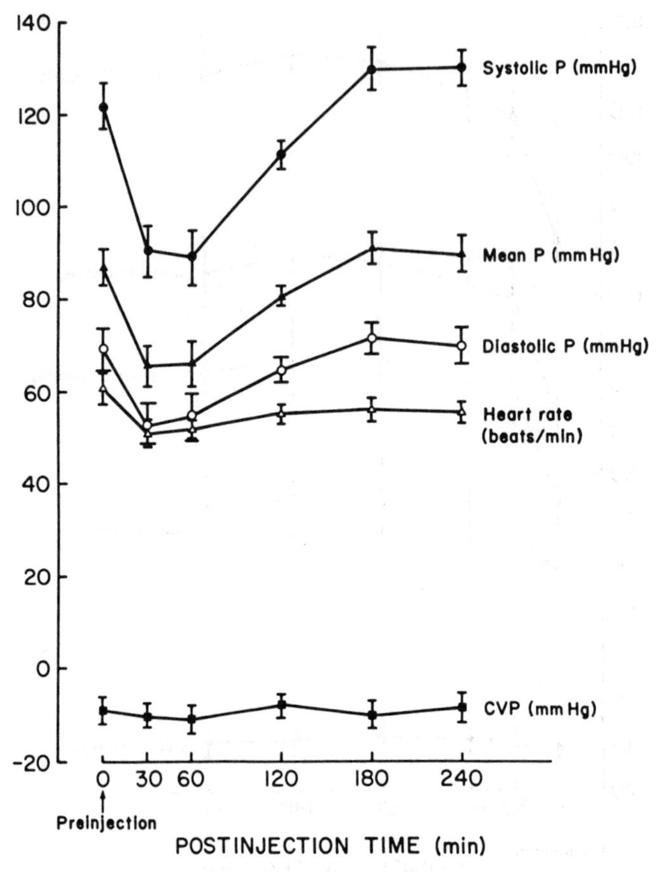

Fig. 6. Changes in heart rate and blood pressures after xylazine administration (2.2 mg/kg). Each point represents mean ±SEM.

for behavioral changes after drug administration. Sedation, restlessness, and muscular shivering occurred in some. Restless ponies exhibited pacing, pawing, body swinging, and/or head shaking.

After butorphanol administration, two ponies did not show any effects. Four were restless at 15 min postinjection with an average duration of action of 1.5 h. One pony was sedated at 20 min and recovered 40 min later. Shivering was observed in one pony at 5 min postinjection and lasted ~2 h. Twenty minutes after butorphanol, one pony was sedated for 2 h before showing shivering, which lasted beyond the final observation at 4 h.

No behavioral effects were observed after flunixin or, in five ponies, after levorphanol administration. Levorphanol induced shivering at 15 min in one pony, which lasted ~1.5 h. Thirty minutes after levorphanol, one pony was sedated for 1.5 h. One pony started pacing 3 h after levorphanol, and the effect continued beyond the final observation at 4 h.

With morphine, one pony showed no effects, while seven were restless. The onset and duration of restlessness were ~1 and 3 h, respectively. Xylazine produced sedative effects in six ponies with the onset and duration of action ~15 and 105 min, respectively. Two ponies given xylazine did not show any behavioral changes.

Experiment IV

Materials and methods

Six healthy adult horses of mixed breeds, two males and four females, weighing from 338 to 410 kg and ranging from 4 to 20 yr of age were used. Substances tested were: butorphanol tartrate, 0.05, 0.10, 0.20, and 0.40 mg/kg; pentazocine lactate, 2.2 mg/kg; and placebo, 0.04 ml/kg. Placebo was the liquid vehicle for butorphanol.

Drug evaluation was performed according to a six-by-six Latin square no less than 3 wk after surgical preparation of the animals. Superficial and visceral pain were induced as in experiment III. Observations were conducted before drug injection (daily control value), after injection at 15 and 30 min, then at 30-min intervals until 4 h postinjection. The tested agent was administered intravenously over a 1-min period. A washout period of no less than 3 days was allowed between drug treatments.

Student's *t* test was used to analyze the data. The individual relative changes of pain threshold [postinjection threshold − preinjection threshold (daily control)/preinjection threshold] after drug injection were used to compare analgesic effects of the placebo, pentazocine, and various doses of butorphanol at each time interval. Significant difference ($P < 0.10$) of postinjection pain threshold from preinjection threshold indicated the test drug had an analgesic effect.

Results

Butorphanol doses of 0.05, 0.10, 0.20, and 0.40 mg/kg are referred to as 0.05, 0.10, 0.20, and 0.40 butorphanol, respectively.

Analgesic effect. No significant analgesic effect was observed for superficial pain after placebo, the negative control, was administered (Fig. 7). After placebo injection, the pain threshold decreased from beginning to end of the test period. Significant analgesic effect produced by pentazocine, the positive control, was observed at 15 min postinjection. The 0.05 butorphanol produced detectable analgesia at

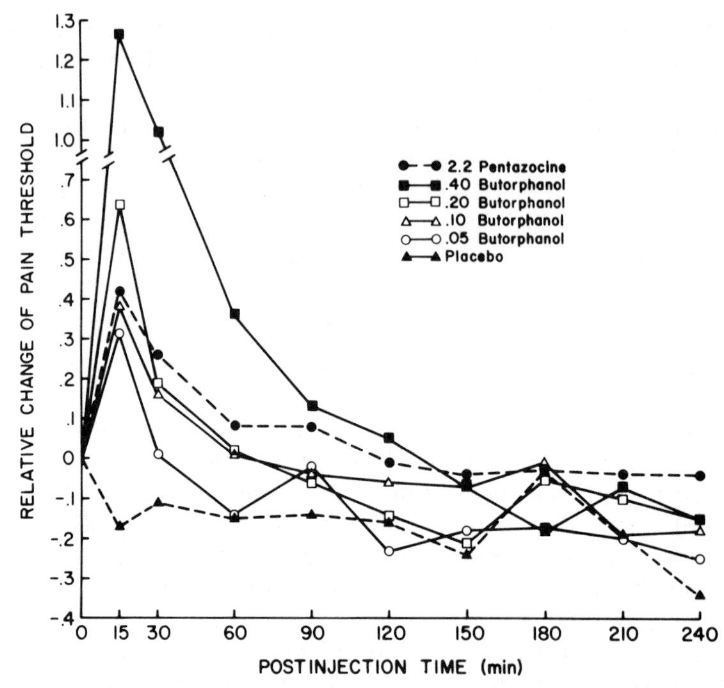

Fig. 7. Relative changes in superficial pain threshold after drug administration. Each point represents mean value.

15 min; after 0.10, effects were seen at 15 and 30 min. Analgesia was detected at 15 min with 0.20 butorphanol, whereas 0.40 was effective at 15 and 30 min.

When compared with preinjection values, neither pentazocine nor any dose level of butorphanol produced significant analgesia for superficial pain after 30 min. Therefore comparison among the tested agents was performed within this period. Pentazocine and 0.05 butorphanol were better than placebo at 15 min. Although 0.10 was better than placebo at 15 and 30 min, analgesia from 0.20 was observed only at 15 min. The 0.40 butorphanol was better than placebo at 15 and 30 min.

Four doses of butorphanol were compared with pentazocine in relieving superficial pain. There were no significant differences between pentazocine and 0.05, 0.10, or 0.20 butorphanol. The 0.40 was better than pentazocine at 15 min.

The effects within 30 min of the four dose levels of butorphanol were also compared. Significant differences were not observed between 0.40 and 0.20 or between 0.20 and 0.10. The 0.40 dose was better than 0.05 or 0.10 at 15 and 30 min. Although 0.10 was better than 0.05 at 30 min, superior effects of 0.20 over 0.05 were not detected.

When compared with preinjection threshold, placebo (negative control) did not have an analgesic effect for visceral pain (Fig. 8). In contrast, pentazocine (positive control) produced significant analgesic effects at 15 and 30 min. The 0.05 and 0.20 butorphanol had effects at 15 min. No difference from preinjection threshold was observed after 0.10 injection. The 0.40 butorphanol produced analgesia at 15, 30, 60, 90, and 120 min.

Because the analgesic effects of pentazocine and various doses of butorphanol were not seen after 120 min, comparison among the tested agents was conducted within this time limit. Pentazocine was better than placebo at 15 and 30 min, whereas significant effect of 0.05 butorphanol over placebo was only seen at 15 min. The 0.10 dose showed effect at 30 min. After 0.20, superior effects were observed at 15 min. The 0.40 butorphanol was better than placebo at 15, 30, 60, 90, and 120 min.

Analgesia produced by 0.40 butorphanol was better than that produced by pentazocine at 30, 60, 90, and 120 min. Pentazocine was better than 0.05 butorphanol at 15 and 30 min and better than 0.10 at 15 min. No significant differences between pentazocine and 0.20 butorphanol were observed.

In comparing butorphanol doses, analgesic effects of 0.40 butorphanol were better than those of 0.05 at 15, 30, 60, 90, and 120 min.

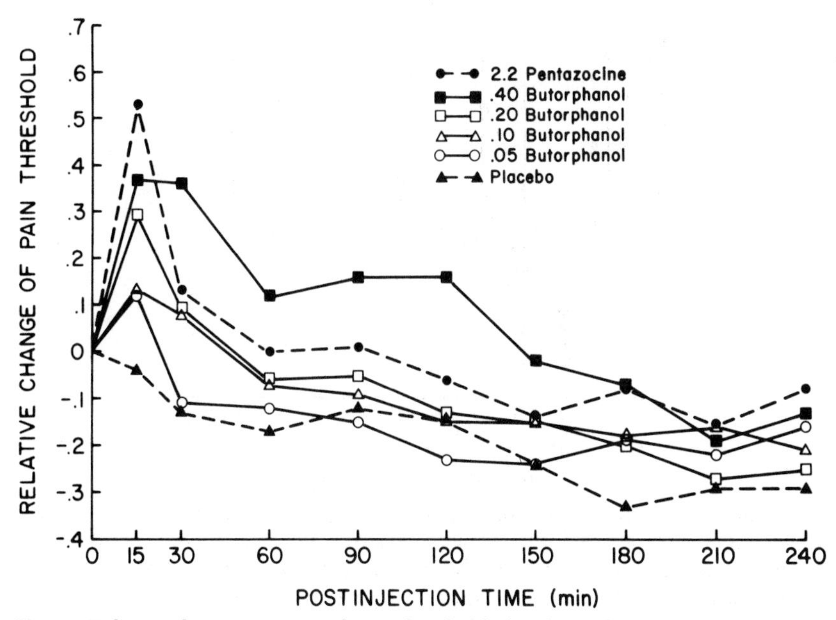

Fig. 8. Relative changes in visceral pain threshold after drug administration. Each point represents mean value.

The 0.40 butorphanol was better than 0.10 at 15, 30, 90, and 120 min and was also better than 0.20 at 60, 90, and 120 min. The 0.20 butorphanol was better than 0.05 at 15 min but did not differ statistically from 0.10.

Based on these data a comparison between the analgesic effects of butorphanol and pentazocine was derived. For superficial pain, analgesic effects of 2.2 mg/kg of pentazocine at 15, 30, and 60 min were equivalent to those of 0.11, 0.13, and 0.14 mg/kg of butorphanol, respectively (Fig. 9). For visceral pain, no dose of butorphanol produced analgesia at 15 min comparable to that of pentazocine (Fig. 10). Anal-

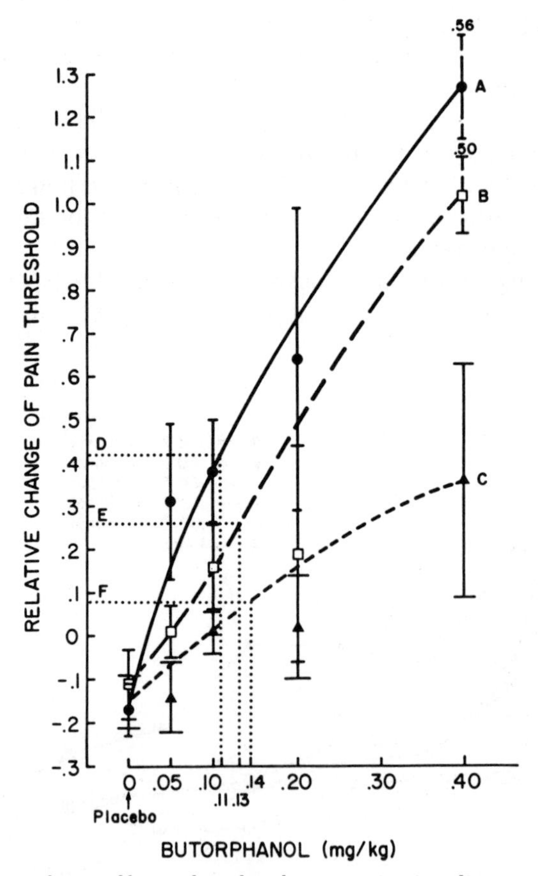

Fig. 9. Dose equivalence of butorphanol and pentazocine in relieving superficial pain. *Curves A, B,* and *C* represent relative changes in pain threshold at 15, 30, and 60 min after butorphanol administration. Relative changes at 15 (D), 30 (E), and 60 (F) min after pentazocine administration were equivalent to changes produced by 0.11, 0.13, and 0.14 mg/kg of butorphanol.

gesic effect of pentazocine (2.2 mg/kg) was equivalent to that of butorphanol (0.22 mg/kg) at 30 min. Pentazocine did not produce analgesia at 60 min; therefore no comparison was made between drugs at this interval.

Behavioral effects. Ataxia, restlessness, shivering, and sedation followed by restlessness were observed in some horses after drug administration. No effects were seen after placebo injection. Immediately after pentazocine was given, all six horses were ataxic and restless for ~80 min, and five horses shivered for a mean of 55 min. A few minutes after 0.05 butorphanol injection, ataxia appeared in two horses and restlessness was seen in four. Approximate mean durations of ataxia and restlessness were 30 and 20 min, respectively. Shivering for 10 min was observed in one horse after 0.05 butorphanol. One horse did not show any behavioral effects.

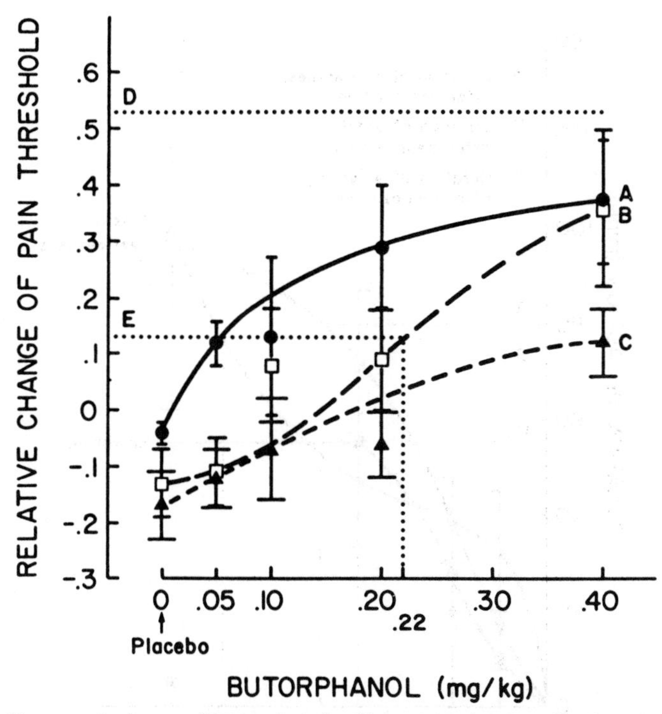

Fig. 10. Dose equivalence of butorphanol and pentazocine in relieving visceral pain. *Curves A, B,* and *C* represent relative changes in pain threshold at 15, 30, and 60 min after butorphanol administration. No dose of butorphanol was comparable with pentazocine at 15 min (D). At 30 min, relative change produced by pentazocine (E) was equivalent to that produced by 0.22 mg/kg of butorphanol. Pentazocine did not change pain threshold at 60 min.

Within the first 10 min after 0.10 butorphanol one horse became ataxic, which lasted 20 min. All six horses became restless; however, two were sedated prior to its onset. Mean durations of ataxia, restlessness, shivering, and sedation before restlessness were ~20, 40, 30, and 20 min, respectively.

Five minutes after 0.20 butorphanol, five horses were ataxic for 45 min. Six horses showed restlessness in ~15 min, which lasted ~70 min. One of the six horses was sedated for 60 min before restlessness. Two horses shivered for a mean of 55 min.

Within the first 5 min, 0.40 butorphanol produced ataxia in three horses and restlessness in four, lasting a mean of 100 min. One horse shivered for 5 min. One horse tolerated this dose of butorphanol without any side effects.

Based on these data a comparison between the behavioral effects of butorphanol and pentazocine was derived (Fig. 11). The durations of

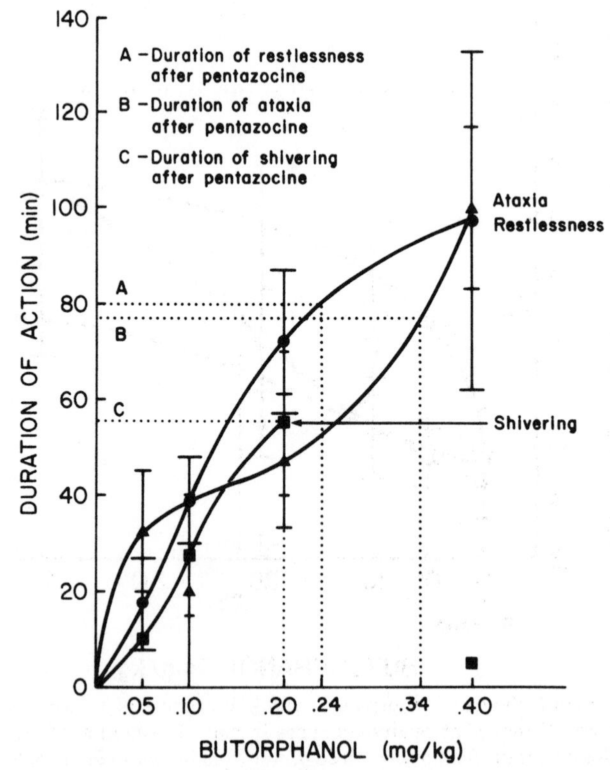

Fig. 11. Behavioral effects of butorphanol and pentazocine in 6 horses. Durations of effects produced by pentazocine (2.2 mg/kg) were equivalent to those induced by 0.20 (shivering), 0.24 (restlessness), and 0.34 (ataxia) mg/kg of butorphanol.

effects produced by pentazocine (2.2 mg/kg) were equivalent to those induced by 0.20 (shivering), 0.24 (restlessness), and 0.34 (ataxia) mg/kg of butorphanol.

Discussion

In these experiments, a Latin square was used to minimize the phenomenon of enzyme induction. Every drug was used one time without prior administration of another drug. In analysis of the results, differences were not found concerning order of administration or days on which drugs or combinations of drugs were used. If enzyme induction occurred, it was not a significant factor.

Xylazine had the best performance in experiment I for relief of three types of pain but was significantly better only in relief of deep pain ($P < 0.05$). This confirmed a subjective evaluation in which xylazine appeared to be the best drug tested for relief of all three kinds of pain. In experiment II the combination of xylazine and fentanyl was significantly better ($P < 0.10$) for relief of visceral pain. This combination was also better than xylazine alone when used for relief of other kinds of pain but was statistically insignificant. This was probably due to an additive action with these two drugs. Klein and Baetjer (32) also found the association of xylazine with morphine superior to xylazine used alone.

Experiment III indicates that xylazine is the best agent for superficial and visceral pain in horses. Morphine could provide analgesia for superficial pain, but more work is needed to prove its efficacy and proper dose level. Butorphanol produced good analgesia for visceral pain and inconsistent analgesia for superficial pain. Because of side effects observed in experiments III and IV, it may not be applicable as a sole agent at high dose levels. A combination of butorphanol and xylazine might improve the analgesic effect of both agents and minimize some of the side effects. Levorphanol appeared to produce only moderate long-term analgesia for visceral pain, but the dose selected might not be optimal. Therefore more investigation is needed before a final decision on its use in horses can be made. Flunixin had no value in alleviating acute superficial or visceral pain, despite the effectiveness reported clinically. This was probably due to the short duration of the stimuli, with little tissue damage and prostaglandin release.

In most human experiments the threshold value is the lowest reaction value obtained. In experiments I and II a mean of all control measurements at 2- and 4-h intervals was used as the threshold value for comparison with drug values at the same intervals, which made those values that varied even on the same day uniform. Making the lowest control value the threshold would have improved the performance of all drugs and drug combinations tested, giving significance to values that were not significant in these two experiments.

High variability in values obtained for each drug for the relief of three kinds of pain could be explained by the variability observed in control values. Variability in the pain threshold was also found by other researchers. The fact that control threshold values declined from day to day can be attributed to sensitization produced by daily testing.

Use of the Pippi pain model represents a step toward objective evaluation of analgesic agents in the equine species. Prior studies have been based almost entirely on subjective evaluation of the animal's response to painful stimuli.

These experiments reaffirm that the equine species is resistant to many of the common analgesic agents. Our work shows that opiates vary in efficacy according to the type of pain stimulus, and virtually all have undesirable side effects. Although xylazine proved best for superficial, deep, and visceral pain, there are many anecdotal reports of horses becoming manic after its administration. Because none of the drugs tested proved to be uniformly effective in all three types of pain, the search must continue for an ideal analgesic agent or combination of agents for the horse.

There is also a need for improved equine pain models. The Tobin behavioral model does not test analgesia. The Pippi model has certain inherent drawbacks. The Brunson model has yet to be shown to correlate with somatic and visceral types of pain.

Our research was supported in part by the Ministry of Education and Culture, Brasilia, Brazil; the Anundhamahidol Foundation, Bangkok, Thailand; and Bristol Veterinary Products, Division of Bristol-Myers Company, Syracuse, New York.

REFERENCES

1. Beecher, H. K. The measurement of pain: prototype for the quantitative study of subjective responses. *Pharm. Rev.* 9: 59–209, 1957.
2. Booth, N. H. Neuroleptanalgesics, narcotic analgesics, and analgesic antagonists. In: *Veterinary Pharmacology and Therapeutics* (4th ed.). Ames: Iowa State Univ. Press, 1977, p. 318–350.
3. Brown, A. K. Levorphan in anaesthesia. *Br. Med. J.* 2: 967–969, 1954.
4. Brunson, D. V., M. A. Collier, and E. A. Scott. Dental dolorimetry as a method for pain threshold studies in the equine species. *Annu. Sci. Meet. Am. Assoc. Vet. Anesth., 2nd, New Orleans, October 1981.*
5. Casy, A. F. The structure of narcotic analgesic drugs. In: *Narcotic Drugs: Biochemical Pharmacology.* New York: Plenum, 1971, p. 1–16.
6. Chapman, W. P., and C. M. Jones. Variations in cutaneous and visceral pain sensitivity in normal subjects. *J. Clin. Invest.* 23: 81–91, 1944.
7. Charpentier, J. Analyse de l'action de la morphine chez le rat par une nouvelle methode quantitative. *C. R. Acad. Sci. Ser. D* 255: 2285–2287, 1962.

8. Ciofalo, V. V., M. B. Latranyi, J. B. Patel, and R. I. Taber. Flunixin meglumine: a non-narcotic analgesic. *J. Pharmacol. Exp. Ther.* 200: 501–507, 1977.
9. Clarke, K. W., and L. W. Hall. Xylazine—a new sedative for horses and cattle. *Vet. Rec.* 85: 512–517, 1969.
10. Combie, J., J. Dougherty, E. Nugent, and T. Tobin. The pharmacology of narcotic analgesics in the horse. IV. Dose and time response relationships for behavioral responses of morphine, meperidine, pentazocine, anileridine, methadone, and hydromorphone. *J. Equine Med. Surg.* 3: 377–385, 1979.
11. Crago, T. G. The use of Innovar-vet injection in canine cesarean sections. *Pract. Vet.* 44: 5, 1972.
12. Davis, L. E., and B. L. Strum. Drug effects and plasma concentrations of pentazocine in domestic animals. *Am. J. Vet. Res.* 31: 1631–1635, 1970.
13. Del Pizzo, A. Butorphanol, a new intravenous analgesic: double-blind comparison with morphine sulfate in postoperative patients with moderate or severe pain. *Curr. Ther. Res. Clin. Exp.* 20: 221–232.
14. Dobkin, A. B., S. Eamkaow, and F. S. Caruso. Butorphanol and pentazocine in patients with severe postoperative pain. *Clin. Pharmacol. Ther.* 18: 547–553, 1975.
15. Dobkin, A. B., S. Eamkaow, S. Zak, and F. S. Caruso. Butorphanol: a double-blind evaluation in postoperative patients with moderate or severe pain. *Can. Anaesth. Soc. J.* 21: 600–610, 1974.
16. Edy, N. B., H. Halbacah, and O. J. Braenden. Synthetic substances with morphine-like effect: clinical experience, potency, side effects, addiction liability. *Bull. WHO* 17: 569–573, 1957.
17. Emmons, M. S., F. B. Coslin, M. W. Safley, and R. T. Tidrick. Use of levorphan tartrate for relief of postoperative pain: preliminary report. *Lancet* 75: 29–31, 1955.
18. Gaard, R. C. Postoperative use of levo-dromoran tartrate. *Lancet* 75: 309–310, 1955.
19. Ganong, W. F. *Review of Medical Physiology* (8th ed.). Los Altos, CA: Lange Med., 1977.
20. Gardocki, J. F., and J. Yelnosky. A study of some of the pharmacologic actions of fentanyl citrate. *Toxicol. Appl. Pharmacol.* 6: 48–62, 1964.
21. Goetzl, F. R., D. Y. Burrill, and A. C. Ivy. A critical analysis of algesimetric methods with suggestions for a useful procedure. *Q. Bull. Northwest. Univ. Med. Sch.* 17: 280–291, 1943.
22. Hall, L. W. *Wright's Veterinary Anaesthesia and Analgesia* (7th ed.). London: Bailliere, Tindall, & Cox, 1971, p. 151–153.
23. Hardy, J. D., H. G. Wolff, and H. Goodell. *Pain Sensations and Reactions.* New York: Hafner, 1967.
24. Hardy, J. D., H. G. Wolff, and H. Goodell. Studies on pain—a new method for measuring pain threshold: observations on spatial summation of pain. *J. Clin. Invest.* 19: 649–657, 1940.
25. Harris, L. S. Structure-activity relationships. In: *Narcotic Drugs: Biochemical Pharmacology*, edited by D. H. Clouet. New York: Plenum, 1971, p. 89–98.
26. Hoffman, P. E. Clinical evaluation of xylazine as a chemical restraining agent, sedative, and analgesic in horses. *J. Am. Vet. Med. Assoc.* 164: 42–45, 1974.
27. Houdeshell, J. W., and P. W. Hennessey. A new nonsteroidal, anti-inflammatory analgesic for horses. *J. Equine Med. Surg.* 1: 57–63, 1977.
28. Hsu, W. H. Xylazine-induced depression and its antagonism by alpha adrenergic blocking agents. *J. Pharmacol. Exp. Ther.* 218: 188–192, 1981.
29. Huber, D. E. A useful clinic aid. *J. Am. Anim. Hosp. Assoc.* 7: 256–258, 1971.
30. Hunt, R. D., and F. F. Foldes. The use of levo-dromoran tartrate (levorphan tartrate) for relief of postoperative pain. *N. Engl. J. Med.* 248: 803–805, 1953.

31. Jaffe, J. H., and W. R. Martin. Opioid analgesics and antagonists. In: *The Pharmacological Basis of Therapeutics* (6th ed.), edited by A. G. Gilman, L. S. Goodman, and A. Gilman. New York: Macmillan, 1980, p. 494–534.
32. Klein, L. V., and C. Baetjer. Preliminary report: xylazine and morphine sedation in horses. *Vet. Anesth.* 1: 2–6, 1974.
33. Krahwinkel, D. J., D. C. Sawyer, and A. T. Evans. Neuroleptanalgesia and neuroleptanesthesia. *J. Am. Anim. Hosp. Assoc.* 8: 368–370, 1972.
34. Kerr, D. D., E. W. Jones, K. Huggins, and W. C. Edwards. Sedative and other effects of xylazine given intravenously to horses. *Am. J. Vet. Res.* 33: 525–532, 1972.
35. Lahoda, R., G. Stacher, and P. Bauer. Experimentally induced pain: measurement of pain threshold and pain tolerance using a new apparatus for electrical stimulation of the skin. *Int. J. Clin. Pharmacol. Biopharm.* 15: 51–56, 1977.
36. Lasagna, L. The clinical evaluation of morphine and its substitutes as analgesics. *Pharmacol. Rev.* 16: 47–83, 1964.
37. Laubie, M., H. Schmitt, and M. Drovillat. Central sites and mechanisms of the hypotensive and bradycardic effects of the narcotic analgesic agent fentanyl. *Naunyn-Schmiedeberg's Arch. Pharmacol.* 269: 255–261, 1977.
38. Leek, B. F. Reticulo-ruminal mechanoreceptors in sheep. *J. Physiol. London* 202: 585–609, 1969.
39. Leek, B. F. Abdominal and pelvic visceral receptors. *Br. Med. Bull.* 33: 163–168, 1977.
40. Lowe, J. E. Pentazocine (Talwin-V) for the relief of abdominal pain in ponies—a comparative evaluation with description of a colic model for analgesia evaluation. *Proc. Am. Assoc. Equine Pract.* 15: 31–46, 1969.
41. Lumb, W. V., and E. W. Jones. *Veterinary Anesthesia.* Philadelphia, PA: Lea & Febiger, 1973.
42. Mackenzie, G. The effect of various tranquilizers in combination with etorphine to produce neuroleptanalgesia in ponies. *Proc. Assoc. Vet. Anaesth. Gr. Br. Ire.* 5: 40–49, 1975.
43. Maxwell, D. R., H. T. Palmer, and R. W. Ryall. A comparison of the analgesic and some other central properties of methotrimeprazine and morphine. *Arch. Int. Pharmacodyn. Ther.* 132: 60–73, 1961.
44. McCarty, J. D. An effective restraint in clinical procedures. *J. Am. Anim. Hosp. Assoc.* 6: 43–44, 1970.
45. McCashin, F. B., and A. A. Gabel. Rompun—a new sedative with analgesic properties. In: *Proc. Annu. Meet. Am. Assoc. Equine Pract., 17th*, 1971, p. 111–118.
46. McCashin, F. B., and A. A. Gabel. Evaluation of xylazine as a sedative and preanesthetic agent in horses. *Am. J. Vet. Res.* 36: 1421–1429, 1975.
47. Miller, L. C. A critique of analgesic testing methods. *Ann. NY Acad. Sci.* 51: 34–50, 1948.
48. Nilsen, P. L. Studies on algesimetry by electrical stimulation of the mouse tail. *Acta Pharmacol. Toxicol.* 18: 10–22, 1961.
49. Nytch, T. F. Clinical observations on the preanesthetic use of oxymorphone and its antagonist, N-allyl-noroxymorphone, in dogs. *J. Am. Vet. Med. Assoc.* 145: 127–131, 1964.
50. Palminteri, A. Oxymorphone, an effective analgesic in dogs and cats. *J. Am. Vet. Med. Assoc.* 143: 160–163, 1963.
50a. Pippi, N. L., and W. V. Lumb. Objective tests of analgesic drugs in ponies. *Am. J. Vet. Res.* 40: 1082–1086, 1979.
51. Pippi, N. L., W. V. Lumb, S. A. G. Fialho, and R. J. Scott. A model for evaluating pain in ponies. *J. Equine Med. Surg.* 3: 430–435, 1979.

52. Pircio, A. W., J. A. Gylys, R. L. Cavanagh, J. P. Buyniski, and M. E. Bierwagen. The pharmacology of butorphanol, a 3,14-dihydroxymorphinan narcotic antagonist analgesic. *Arch. Int. Pharmacodyn. Ther.* 220: 231–257, 1976.
53. Reed, S. M., and W. M. Bayly. Medical management of acute abdominal crises. *Mod. Vet. Pract.* 61: 543–546, 1980.
54. Schauffler, A. F. Acetylpromazine + methadone = better equine restraint. *Mod. Vet. Pract.* 50: 46–49, 1969.
55. Soma, L. R. *Textbook of Veterinary Anesthesia.* Baltimore, MD: Williams & Wilkins, 1971, p. 121–128.
56. Tavakoli, M., G. Corssen, and F. S. Caruso. Butorphanol and morphine: a double-blind comparison of their parenteral analgesic activity. *Anesth. Analg. Cleveland* 55: 394–401, 1976.
57. Tobin, T. Pharmacology review: narcotic analgesics and the opiate receptor in the horse. *J. Equine Med. Surg.* 2: 397–399, 1978.
58. Vernimb, G. D., and P. W. Hennessey. Clinical studies on flunixin meglumine in the treatment of equine colic. *J. Equine Med. Surg.* 1: 111–116, 1977.

TWELVE

Control of Pain in Dogs and Cats

Howard C. Hughes
C. Max Lang
Department of Comparative Medicine, Milton S. Hershey Medical Center, Pennsylvania State University, Hershey, Pennsylvania

The perception of pain is an extremely complex physiological phenomenon, the reaction possibly involving many systems of the body (14). Just how one particular species may respond to a noxious stimulus can be exhibited in different ways. There can also be a great deal of interspecies and individual variability so that one animal may respond quite differently than another. This variability among animals is decreased in some of the common laboratory animals, such as rats and mice, where generations of breeding for laboratory use have tended to produce animals that respond quite uniformly. Variability, however, is often accentuated in highly outbred populations such as dogs and cats. Some breeds of dogs, such as hounds, are noted for their hardiness and may be less likely to exhibit overt signs of pain to the same stimuli, whereas other breeds, such as poodles or pomeranians, may be more sensitive. Cats may only show visible signs of pain when the stress becomes more pronounced. Apparently, therefore, it is the tolerance to pain that is highly variable, even though pain thresholds are remarkably similar among animal species (12).

It is extremely important to recognize the signs of pain in the animals used in research. Pain may be expressed by obscure signs, such as inappetence, lassitude, and dysuria, or by more obvious signs, such as lameness or biting and scratching at an irritation. Sometimes it is only possible to determine if pain is present and how severe it is by knowing how the animal behaved prior to the onset of the painful stimuli.

It may be helpful to evaluate pain in animals by recalling personal experience. By knowing what the inciting stimulus is, how we perceive it, and the behavioral patterns of the species we are working with, one can judge better when pain-relieving measures are necessary. Even though an animal's response to pain is remarkably similar to human response, it is not always valid to transfer the degree of severity to the animal. Animals do not have the behavioral response to pain often found in humans. For example, the response of dogs and cats to operative procedures such as a hysterectomy or herniorrhaphy appears to be much less than the human response to similar procedures. Animals are often active the day of surgery and within a few days they are running and behaving as though nothing had happened. With the same procedure and at the same point postoperatively, adult humans may just be able to get out of bed. This difference is not because veterinary surgeons are better than human surgeons, but rather because of years of learned behavior in an adult human, for whom anticipation and suggestion may become as important for the perception of pain as the operation itself. Animals may have a higher pain threshold than humans undergoing similar procedures (3). Therefore it is important to temper judgment of a painful experience with careful observations of animals and the individual response to the stimuli.

When the noxious stimulus is indeed painful, it is necessary to evaluate the animal's condition to determine what pain-relieving measures, if any, should be taken. Pain is a sign of injury and acts to restrict activity that aids in healing. For example, muscle and joint pains can often be quickly and effectively eliminated by the use of anti-inflammatory drugs, but these agents may make the patient more active and thus compound the injury. Other factors to consider include drug-related side effects such as depression of the cardiovascular, respiratory, or hematopoietic systems. There are no risk-free agents effective against severe pain.

A pain reliever should only be selected after determining that the animal can, in fact, be made more comfortable by the use of pain-relieving drugs and that these agents will not injure the animal or allow for the potentiation of an existing injury. In research it may also be necessary to determine if these agents will interfere with the study. Therefore the investigator should be informed of the actions, distribution, and metabolism as well as any interactions of agents to be used. Some of these factors, as well as the basic pharmacology that may facilitate the selection of one of the more commonly used drugs for use in dogs and cats, are reviewed in this chapter.

Narcotics

Morphine

Narcotics are some of the most widely used and potent therapeutic agents for the relief of pain. Morphine was the first analgesic to be purified, and, although many other agents have since been synthesized, all other drugs are still compared to it (6). Morphine has an effect on many of the organ systems of the body, but it is used primarily because of its effects on the central nervous system (CNS) and gastrointestinal tract (5). Although classically thought of as a CNS depressant in humans, its effects can be quite irregular in different species. In cats, morphine at higher doses may be associated with severe excitement, tonic spasms, and aggressive behavior (2) and was long thought to be contraindicated. In dogs, there is a brief period of excitement characterized by restlessness and salivation that is shortly followed by nausea, vomiting, and defecation. Once these signs are past, however, the dog becomes much less responsive to pain and may progress into a stuporlike sleep at higher doses.

When used as an analgesic, morphine has the ability to decrease pain without inhibiting motor activity (5). This effect may not be due to an actual elevation in the patient's pain threshold but rather to an increase in the ability to tolerate the pain. The patient may be able to perceive the stimulus, but the fear, anxiety, and panic associated with the pain are significantly reduced by morphine.

Morphine's effects on the gastrointestinal tract are twofold (6). First, it induces nausea, vomiting, and defecation by direct stimulation of the chemoreceptor trigger zone in the medulla. After the initial stimulation, there is an overall slowing of the gastrointestinal system that is more or less proportional to the dose.

The other actions of morphine are primarily related to the respiratory and cardiovascular systems. Morphine rapidly depresses respiration by a direct effect on the brain. This effect is evident even with a subsedative dose and increases in severity with increased doses. Morphine affects all phases of the respiratory cycle, including rate, minute volume, and tidal volume, with resultant hypercapnia.

The effects on the cardiovascular system are minimal. A decrease in blood pressure and heart rate, like that seen with sleep, is probably more closely related to the animal's reduced activity and sleeplike state than to a direct effect on the cardiovascular system.

Profound analgesia and sedation can be produced by a single 2–5 mg/kg intramuscular or subcutaneous inoculation of morphine sulfate for the dog and 0.1–1.0 mg/kg subcutaneously for the cat. The usual analgesic dose is 0.25 mg/kg for the dog and 0.1 mg/kg for the cat. Maximal therapeutic effects are seen in ~1 h and last 6–8 h. About 65% of the dose is excreted within 24 h. To reduce many of the early gastrointestinal stimulatory effects of morphine, an anticholinergic agent, such as atropine sulfate (0.05–0.1 mg/kg), will serve to minimize the discomforting nausea, retching, and vomiting (1, 2).

Meperidine

Meperidine (Demerol) is a synthetic narcotic analgesic that is free of many of morphine's undesirable characteristics. Its major site of action is on the CNS, where it acts much like morphine. The onset of effects is rapid (<10 min), but its duration of analgesia is only 45 min in the dog and cat. Meperidine effects on the cardiovascular, respiratory, and gastrointestinal systems are similar to those of morphine; however, meperidine does not stimulate the chemoreceptors in the brain stem and therefore does not induce nausea, vomiting, or defecation (1, 5).

Meperidine is an effective analgesic, although less potent than morphine. It also has a somewhat shorter duration of action and therefore must be given more frequently. Meperidine is safe for both dogs and cats. A dosage of 2–10 mg/kg intramuscularly is satisfactory for pain relief. This dosage can be repeated safely 2–3 times daily without fear of overdosing. Meperidine's CNS-stimulating properties are not seen until the dosage exceeds 25–30 mg/kg (7).

Fentanyl

Fentanyl is a relatively new narcotic analgesic that is 80–100 times more potent than morphine (10). Its use in veterinary medicine has been mostly confined to the dog, and it is generally combined within the neuroleptic tranquilizer droperidol in a preparation called Innovar-Vet. The onset of action of fentanyl is seen within 3 min after intravenous and 10–15 min after intramuscular injection. Its duration of action is short, lasting only 30–60 min, and therefore its usefulness is directed more toward surgical than postoperative analgesia. When used in conjunction with local or general anesthetics, fentanyl provides the powerful analgesia that is often missing with some of the general-anesthetic agents. This enables a decrease in the dosage of general anesthetics for minor procedures or in the critically ill patient.

Two notable adverse effects produced by fentanyl are profound bradycardia and sensitivity to external noise. This auditory sensitivity seems to be peculiar to the dog and can be quite annoying because sudden or inadvertent noises in the operating room produce arousal of the animal. This effect can be minimized if Innovar-Vet is used as an analgesic in conjunction with a general anesthetic.

There may also be profuse salivation and defecation with Innovar-Vet, which, like the bradycardia, is probably a result of parasympathetic stimulation. Bradycardia, salivation, and defecation may be minimized by the use of atropine sulfate (0.05 mg/kg), which is most often recommended. Innovar-Vet is not recommended for use in cats because it produces CNS stimulation rather than depression.

Fentanyl (0.4 mg/ml) is supplied only in combination with droperidol (20 mg/ml) as Innovar-Vet for use in dogs. The recommended intramuscular dosage is 1 ml/7–10 kg, or it may be given slowly intravenously at 1 ml/10–20 kg.

Antagonists

It should always be kept in mind that all narcotics are habit-forming and are controlled substances subject to abuse. In addition the prominent feature of respiratory depression, although sometimes due to inadvertent overdosing, has limited the clinical usefulness of narcotics for long-term analgesics in veterinary medicine.

Narcotic antagonists should always be kept available when the opiate derivatives are used. Naloxone and nalorphine are the most prominent of these drugs (1). These agents inhibit the binding of the opiates at receptor sites, which quickly and reliably reverses the respiratory and CNS depression, muscular incoordination, vomiting, bradycardia, and the other adverse side effects of narcotics. Not only are these antagonists useful in an overdose situation, but they should be considered adjuvant to fast, smooth recovery of narcotic preanesthetized or sedated patients. Dosage of the antagonists varies depending on the depth of sedation and type of narcotic used. Generally, an intravenous dose of 0.04 mg/kg naloxone or 1 mg nalorphine/10 mg morphine or 20 mg meperidine should reverse the narcotic effects.

Nonnarcotic Analgesics

The nonnarcotic analgesic drugs relieve mild to moderate pain without causing sleep or sedation. However, these agents are not effective in relieving severe somatic or visceral pain (5). In addition to pain relief, these agents often have some degree of anti-inflammatory and antipyretic action. Nonnarcotic analgesics are believed to act on the brain at the subcortical level probably by increasing the pain threshold. Their antipyresis is produced by inhibition of prostaglandin synthesis in the hypothalamic heat-regulatory centers.

The most commonly used nonnarcotic analgesic drugs are acetylsalicylic acid (aspirin), acetaminophen (Tylenol), and phenacetin (Empirin). These are the over-the-counter analgesics found in most home medicine cabinets. They are supplied as single agents, combination products, or with buffers. There is no evidence, however, to support the assumption that combination drugs have effects that are any more than just additive (5). Such combinations are no better than when the proper doses of single agents are used. In general, their actions are all quite similar, none having particular advantage over the others.

Clinically the analgesic use of these agents should be limited to pain associated with muscles, joints, and peripheral nerves because these analgesics do little for deep visceral-type pain. They are also of value as antipyretic agents reducing fever, but they have no effect on normal body temperature (5, 14).

Adverse side effects from these analgesic agents are probably more commonly seen in dogs and cats than in other species. Because they are available over the counter, the tendency is to use them indiscriminately and at higher-than-therapeutic dose levels. Note also that the cat is much slower than other animals in its rate of hepatic drug metabolism due to a deficiency of bilirubin-glucuronoside glucoronosyltransferase (4). Inadvertent overdosing and individual idiosyncrasies sometimes produce epigastric distress, nausea, and vomiting. Other complications, although infrequent, include renal and/or hepatic insufficiency, aplastic anemia, or thrombocytopenia.

In cats the proper dose of aspirin is 10.5–25 mg/kg (0.15–0.35 grains/kg) once every 24–48 h (13); this dosage is safe and produces effective blood levels without side effects after 2 wk of continuous use. Similar dosage, but divided so that the drugs are given 2 times daily, is equally effective in the dog (8). Buffered or enteric-coated tablets should help alleviate gastric distress if it occurs. Acetaminophen produces severe hemolytic anemia and hepatocellular necrosis in cats and should not be given.

Several other nonnarcotic analgesics are available for use in human beings. Although none are specifically approved for use in animals, they may provide some benefit for short-term analgesia (5).

The first of these is propoxyphene. Alone it is a mild analgesic, but when combined with aspirin, phenacetin, and caffeine (APC) it produces a greater level of analgesia than when any of these agents is used alone. Although pharmacologically related to methadone, propoxyphene does not compare in its production of analgesia or in potential for habituation. It is used for the relief of mild pain much like the other nonnarcotic analgesics. Although no specific dosage is published for dogs and cats, the propoxyphene-APC product can be used with the dosage and schedule of administration of aspirin (227 mg) and phenacetin levels (162 mg).

Pentazocine is a potent analgesic that can be given orally or intramuscularly and, on a 1 mg/1 mg basis, has an analgesic potency equivalent to codeine. The onset of action is rapid, usually 15–20 min, and lasts 30–45 min in the dog. As an indication for dosage, 30 mg of pentazocine is equivalent to 75–100 mg of meperidine. It is used for relief of moderate to severe pain and as a supplement to surgical anesthesia. In people, pentazocine is sometimes accompanied by hallucinations, disorientation, and confusion. A similar response might be seen in cats and dogs.

Anti-Inflammatory Agents

Steroidal and nonsteroidal anti-inflammatory drugs are available for use in dogs and cats. These agents suppress cellular and chemical response to injury by decreasing the activity of fibroblasts, neovascularization, and granulation tissue formation (15). The exact mechanism by which this occurs is not known, but it is probably related to suppression of the cellular response and protein catabolism. These agents are used to alleviate many symptoms, but they are especially useful for arthritis and allergy. They are not curative, however, and symptoms reappear once therapy is discontinued. In addition, they produce a euphoria or feeling of well-being in humans that may be beneficial in terminal conditions such as cancer. Their usefulness in shock is controversial, but they may aid in stabilizing the patient. However, for all the usefulness of corticosteroids, these agents have a wide range of adverse side effects—from increased food and water intake to actual dependency. They are contraindicated in animals with infections or for postoperative pain, because they inhibit the normal healing processes.

Dosage of the corticosteroids depends as much on the agent as on the condition being treated (1). For physiological replacement, hydrocortisone is given at 1–2 mg/kg initially, with maintenance levels as low as 0.1–0.5 mg/kg once or twice daily. Other steroids are given at doses based on their potency compared to hydrocortisone. These agents are effective by intravenous, intramuscular, intra-articular, or oral routes. In chronic conditions, long-term therapy should consist of an alternate-day regime of prednisolone to lessen physical dependency in the form of hypocorticoidism.

The best-known alternative to the corticosteroids is phenylbutazone (Butazoladin) (1, 14). Its anti-inflammatory, analgesic, and antipyretic actions are useful in the treatment of arthritis and skeletal disorders. This agent is not widely used in dogs, probably more because of its reported toxicity in humans than its actual toxicity in dogs. Phenylbutazone has been associated with gastrointestinal, renal, cardiac, and blood disorders. Phenylbutazone is quite toxic in cats even in moderate doses and therefore should be avoided. Dosage in the dog is ~10 mg/kg given three times daily.

Miscellaneous Drugs

Muscle relaxants

Muscle relaxants, when given intravenously or orally, have been found useful in the treatment of acute muscle spasms seen after trauma, disk disease, or inflammation. They also aid muscle relaxation during orthopedic surgery. Many agents that are widely used for muscle pain and tension are also used to relieve anxiety and psychosomatic disorders (5). Much of the beneficial properties may also be due to the sedative effect produced by such agents as diazepam (Valium). Anxiolytic agents are also useful in combination with general anesthetics such as ketamine and pentobarbital. They potentiate the anesthetic while allowing for lower doses, in addition to relieving anxiety and relaxing muscles.

Ataractics

Ataractic or neuroleptic drugs, such as droperidol and acetylpromazine, potentiate narcotics and anesthetics to provide chemical restraint and to relieve anxiety. The tranquilizers fall into major and minor categories. Drugs of the major group, such as phenothiazines, are used for chemical restraint, preanesthesia, and sedation. These have not been shown to alleviate anxiety or pain in nonsedative doses. Drugs of the minor group, or anxiolytic drugs such as the benzodiazepines and propanediol, are used to reduce stress associated with environmental changes or behavioral conditioning. These agents are often effective in alleviating internal stress without noticeable sedation. Because animals can not tell us when their anxieties and fears are gone, as humans can, these drugs are used more as sedatives and behavioral modifiers than anything else. Effects of ataractics on pain may be through their muscle-relaxing properties, or they may actually raise the animal's pain threshold.

Ketamine

One of the newer injectable agents is ketamine hydrochloride (9). It is mentioned here primarily because of its unique ability to produce effects from sedation to deep anesthesia, depending on the dose given and its wide margin of safety. Ketamine is unique in that its lack of cardiorespiratory depression is unsurpassed by any other agent. It can be given intramuscularly or intravenously and has been found compatible with many other drugs. Ketamine provides profound superficial analgesia; however, visceral pain may still be present. The latter can be abolished by the use of other analgesic agents. The actual state produced by ketamine resembles catalepsy. There is a generalized hypertonicity to the muscle, mild nystagmus, cardiovascular stimulation, lack of respiratory depression, and the laryngeal and pharyngeal reflexes remain intact.

Its onset of action is rapid, only 25–30 s after intravenous and 4–5 min after intramuscular administration. The dosage is 2–45 mg/kg in the cat and 10–100 mg/kg in the dog depending on the route of administration and degree of sedation desired (7, 14). The broad therapeutic index and lack of long-term toxicity make ketamine very useful for repeated sedation at close intervals. In addition, its cardiovascular stimulation makes it a good agent for the debilitated or hypovolemic patient. Because ketamine does not produce visceral analgesia, it should be used in combination with other agents when visceral pain is expected.

The principal adverse reaction of ketamine is during recovery. Ketamine produces mild hallucinations, which may be manifested by what appears to be mild convulsions or muscle jerking. It is more prominent in cats than dogs. The reaction is potentiated by outside stimuli. Therefore it is best to keep the animals in a quiet, dark area and avoid poking or otherwise stimulating them during recovery. These effects can also be alleviated by tranquilizers, ultrashort-acting barbiturates, or other anesthetics.

Summary

None of these drugs is free from side effects and many are habituating. It is best to design studies with temporary and minimal pain. Analgesics should be considered when the animal is uncomfortable from its pain. These agents should be used for the short-term relief of pain. When such pain becomes severe or chronic, specific measures directed toward the pain-inciting stimuli should be used.

No single analgesic or anesthetic is suitable for use in all situations. One may be lulled into a false sense of security by successes with a specific method or drug. This is especially true for normal patients, to whom it makes little difference what regimen is used. Too often convenience, habit, ease of administration, or economics determines what agent is used. An agent that works for healthy patients may not be suitable for sick ones, and one that works well for one species may not be equally good for another. One should not study new techniques on sick patients; rather one should learn the techniques associated with the procedure first. New techniques may be plagued with technical errors, and the patient may not be able to tolerate mistakes.

Some species simply do not do well with certain regimens, even if these regimens are well managed. A properly prepared scientist has several techniques available. The pharmacological effects of the drugs and how they interact with normal and abnormal physiological processes in that species are understood. This scientist would also know how to compensate for bad effects of a drug and how to make the most of its good effects. A good regimen must be tailored to fit a specific set of circumstances.

Our study was supported, in part, by Public Health Service Grants R01-RR00469, K04-HL00586, and R01-HL13988.

REFERENCES

1. Booth, N. H. Drugs acting on the central nervous system. In: *Veterinary Pharmacology and Therapeutics*, edited by L. M. Jones, N. H. Booth, and L. E. McDonald. Ames: Iowa State Univ. Press, 1977, p. 191.
2. Bowen, J. M. Drugs acting on the nervous system. In: *Handbook of Laboratory Animal Science*, edited by E. C. Melby, Jr. and N. H. Altman. Cleveland, OH: CRC, 1976, vol. 3, p. 65.
3. Brain, L. Animals and pain. *New Sci.* 19: 380–381, 1963.
4. Brodie, B. B. Of mice, microsomes and man. *Pharmacologist* 6: 12–26, 1964.
5. Goodman, L. S., and A. Gilman. *The Pharmacological Basis of Therapeutics* (5th ed.). New York: Macmillan, 1975.
6. Hall, L. W. *Wright's Veterinary Anesthesia and Analgesic*. Baltimore, MD: Williams & Wilkins, 1971.
7. Hughes, H. C., W. J. White, and C. M. Lang. Guidelines for the use of tranquilizers, anesthetics and analgesics in laboratory animals. *Vet. Anesth.* 2: 19–24, 1975.
8. Latt, R. H. Drug doses for laboratory animals. In: *Handbook of Laboratory Animal Science*, edited by E. C. Melby, Jr. and N. H. Altman. Cleveland, OH: CRC, 1976, vol. 3, p. 561.
9. Lanning, C. F., and M. H. Harmel. Ketamine anesthesia. *Annu. Rev. Med.* 26: 137–141, 1975.
10. Lumb, W. V., and E. W. Jones. *Veterinary Anesthesia*. Philadelphia, PA: Lea & Febiger, 1973.
11. *Occupational Exposure to Waste Anesthetic Gases and Vapors*. Natl. Inst. of Safety and Health, publ. no. 77-140, 1977.
12. Vierck, C. J. Extrapolations from the pain research literature to problems of adequate veterinary care. *J. Am. Vet. Med. Assoc.* 168: 510–514, 1976.
13. Yeary, R. A., and W. Swanson. Aspirin dosages for the cat. *J. Am. Vet. Med. Assoc.* 163: 1177–1178, 1973.
14. Yoxall, A. T. Pain in small animals—its recognition and control. *J. Small Anim. Pract.* 19: 423–438, 1978.
15. Yoxall, A. T. Pain and inflammation: analgesics and anti-inflammatory agents. In: *The Pharmacological Basis of Small Animal Medicine*, edited by A. T. Yoxall and J. E. R. Hird. Oxford, UK: Blackwell, 1978.

Index